voice

Dear Michael. Thanks for Shaing your light
Kristen xoxo

How to Share Your
Message, Your **Products**
and Your **Business** with
the World!

Kristen White

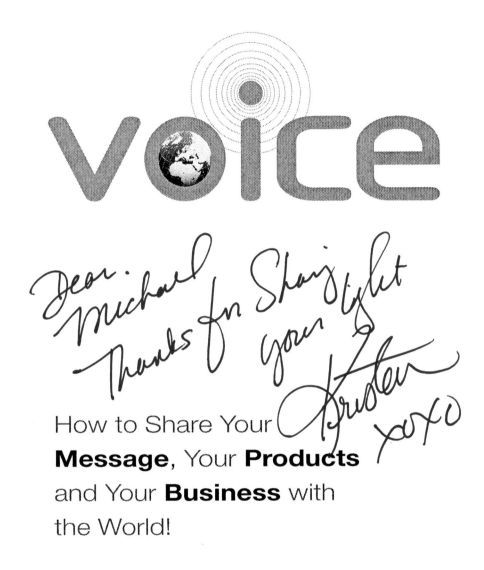

VOICE, How to Share Your Message, Your Products and Your Business with the World!

Published by MysticMedia, LLC
6614 Clayton Rd
Suite 387
St. Louis, MO 63117

www.kristenwhitetv.com

Manufactured in the United States of America.

ISBN: 978-0-9828424-1-6

To my husband who listens and

my kids who don't, and to the many voices

who spoke to me along the way.

And to the long line of those who have

something great to say but have not yet

found their voice.

And to the millions who will be touched tomorrow

*by the **VoiceRipples™** we create today.*

Contents:

"What do you have to say?

Who do you want to be?

What are you destined to do?

Life is about the impact you make

with the message you share.

Find your voice...

Let's start a ripple!"

Kristen White

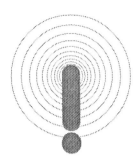

Preface:
The Time is Now!

It's a new world...of potential and opportunity! Television stations, radio stations, and publishing houses across the country are undercapitalized, understaffed, and packed with outdated equipment. Old communication technology is giving way as new media channels are coming on the scene to meet the demands of a world hungry for information and inspiration.

The resulting shift has created a tremendous window of opportunity for those who have a passionate message to share.

The world is filled with over a billion seekers of solutions to soften their suffering—and in this new media world you can rise purposefully and rapidly from anonymity into niche expert status. What you need is strong intent, a compelling message, and a loud-and-clear mental readiness to shift into decisive action via a well-tuned strategy for media exposure and a consistent blueprint for ongoing communication to the community you attract.

Experts can publish, broadcast, promote and publicize their content without restriction. The only obstacle remaining lies within the mindset of the individual and their belief in their ability to step into the spotlight and inspire others with their message.

If your soul is burning with a needed solution that you long to share, then read this book, master the inner preparation and media steps necessary to attain a strong voice and high-visibility platform, and then go forth and share your vision with the world! The time is NOW.

Manifest Destiny

Mankind has always grown through its curious nature. This curiosity has sparked the desire to explore and conquer the unknown. Throughout the centuries, individuals have sailed across oceans, traveled into foreign lands, and laid claim to territories that appeared to belong to no one. This inclination remains strong today. However, on our physical planet there are very few unclaimed territories that are desirable for the creation of cities and communities.

Now I believe the unclaimed territory exists online, or in the "cloud." The cities are no longer physical cities. Instead, they are inspirational cities created by a global thought leader who connects like-minded people who want to dwell in a community of similar ideas, values, beliefs, and practices.

Travel back in time with me for a moment. In the early 1800s, many families and individuals packed all of their belongings to head off into the unknown territory of the western United States to create a new beginning for themselves. These individuals believed that the cities were too crowded, full of crime and competition, and lacked the opportunity for growth and expansion that was present in the unsettled territories out West.

This mass migration was called **Manifest Destiny. In the 1840s, it was a widely held belief that the population of the United States was destined by God to expand from sea to sea.** This popular philosophy did not take into account the indigenous tribes who had lived on the land for thousands of years. These tribes were all without a voice in the discussions about Manifest Destiny. It is a perfect example of why the voice for each individual point of view needs to be present in conversations that involve the collective whole. However, individuals—predominantly immigrants to the U.S.—believed it was their destiny to claim a piece of physical territory in a wild and unexplored place. **Their motivation was a deep emotional desire to own land and create a homestead, giving their family name a new beginning and a lasting legacy based on their ideas and their trade.** Many people on this journey were not successful. Every day on the trail, men, women, and children lost their lives to illness, hostile tribes, criminals, weather, and geographic obstacles. The cost was high, but the emotional investment in the belief was stronger and outweighed any risk, including death.

One hundred years later, the people who were successful are now the founders of some of the largest cities in the West. In many territories, one family name is responsible for the birth of an entire metropolitan area now inhabited by millions of individuals.

Now let's head back to today where we are in a time of manifest destiny online. There are many individuals like you who have ideas that are emerging and have yet to be accepted by the mainstream.

In our modern world, many of these ideas are drowned out by the noise of crowded mainstream media communities where there is limited opportunity for a fresh voice to emerge or the ability to broadcast at a level to be heard. Like the settlers more than 150 years ago, experts have to pack up their wagons with supplies and head for unclaimed territory online. In this context, the wagon is metaphorical: it represents your message, your business, your book. The wheels of the wagon are your ability to move your message forward and to spread it through direct contact with as many listeners as possible. As with all journeys into unclaimed territory, the obstacles will not kill you personally, but they may kill your hopes for the success of your book, product, or business.

Here are the parallels: Illness is represented by your mindset and your inability to give the necessary nourishment to your expert platform so that it can properly grow. Criminals are individuals along the way who will will "borrow" your intellectual property and call it their own. Finally, weather and geographical obstacles are the many unexpected challenges with any project that create setbacks both in terms of time and finances.

Truth be told, the successful settlers all had one thing in common: They took a giant leap of faith. They loaded their wagon with everything they owned. There was no Plan B. They headed out into a wild land and persevered—day in and day out—until they reached a spot where they were inspired to stop traveling and start settling. Often, this land had nothing, not even a house. The settlers built the house, the barn, and settled their territory one stick at a time. They did not give up. They took action. The job required 100% daily focus and determination. The same is true today.

If you want to use your voice to create an expert platform that claims a territory online around which an inspirational city can blossom and grow into a following of millions, the same path as outlined above will be required of you personally. You will need to build it step by step and persist until your voice is heard. Are you up for the task?

It's Time to Load Your Wagon

There are many messengers who have traveled ahead of you and are already magnetizing their communities, yet, there is still plenty of unclaimed territory available online. This will not always be the case. In this very moment there exists an unprecedented opportunity. There are no barriers in existence that restrict your ability to share your voice with the world.

No less than two decades ago, the Internet did not exist—publishing houses were elite gatekeepers, and the media was governed by a few. Now an unknown individual can create an online presence in a matter of days; write, publish, and promote a book showcasing their ideas; and then create a multimedia broadcast of their philosophies that can be heard around the world. This is possible without the interference of culture, religion, and government.

The only obstacle that remains today lies within the mind of the individual: your belief in your ability to share ideas and instructions designed to inspire others. Self-esteem, lack of focus, clarity of message, and fear will stop 99% of you before you start. This book is for the mere 1% of you who have the courage and conviction to travel this path.

Your legacy will live on, your inspiration will endure, and you will be remembered.

"It is from numberless diverse

acts of courage and belief that human

history is shaped.

Each time a man stands up for an ideal,

or acts to improve the lot of others,

or strikes out against injustice,

*he sends forth **a tiny ripple of hope.**"*

Robert F. Kennedy

Introduction:
Why Broadcast
Your Voice?

The **main purpose of this book** is to explore a more effective and inclusive approach for becoming a successful expert in this new millennium. To succeed in today's marketplace you need to prepare yourself to step in front of the camera and go public with your message, product, and service. In this process of using the media to gain visibility, you become an authentic celebrity in your chosen field of expertise.

Specifically, we're going to examine in detail how you can **pre-**

pare your inner presence and message, and then attain a higher visibility online and in the media as an expert, so you generate impressive support for your particular service, vision, or cause.

A media expert is anyone of considerable knowledge in a particular field who develops a special charge of personal charisma, and then steps onto the public or online stage with confidence and purpose.

Traditionally, being a media celebrity was viewed mostly in terms of pretty people gaining access to fancy parties, red carpet events, Hollywood houses, expensive cars—and being covered in slick magazines like *Star* and tabloids like *The Inquirer*. These media personalities appear in movies and television dramas designed to entertain us. These celebrities are a distraction from the harsh realities of our everyday lives. Their entertainment is a powerful, yet fleeting, form of escape from the daily reality of debt, loss, health issues, and personal struggle. While this type of media celebrity still continues unabated, there's now a new definition of celebrity on the rise where one is lauded for being the recognized **enlightened expert** in an area where the world truly needs a permanent shift in insight and help to transcend whatever obstacle life has presented.

When approached properly, becoming an enlightened expert will naturally generate its own form of celebrity within your specific audience, and the attraction that your passion and expertise create can make a strong impact on a national or even global level.

We use the term "enlightened" here to mean the attainment of a new and unique insight into a particular problem that many people encounter. In this spirit, enlightened experts can significantly shift the planetary mindset with their personal wisdom and guidance—and an enlightened celebrity can spark the birth of a vital new movement.

Music and movie stars all too often serve as media distractions from the problems in our own lives. We focus on their drama and celebrity lives as a way of avoiding our own comparative mediocrity or inner pain. But this fixation on traditional stardom is often a form of escape that doesn't really accomplish anything of lasting value in the world.

On the other hand, an enlightened media celebrity is anyone who has:

1. **The courage and capacity to look deeply and purposefully** at core challenges being faced by our world.

2. **Developed a unique solution** to one of these challenges.

3. **Decided to step forward into the limelight** to present and activate their unique solution using multiple media channels to reach as many individuals as possible.

Often these people have had a unique life journey that they've been able to cultivate into deep understanding and expertise. This expertise may be in business, or in social work, or education. It may be in environmental science, or spiritual growth, or any other field where they have advanced a lifetime of experience into a system of genuine value to humanity.

At some point, these people have evolved the translation of their experience and insight into a teachable or marketable body of work—it's ready and poised to be shared and of service in the world.

The enlightened expert wants to make a difference, to have a positive impact. But in order for people to know what the expert has, and to respond to the offering, that expert must rise to the challenge of giving voice to his or her expertise.

They must become clear in their message and emotionally bright, then step into the spotlight with celebrity poise and deliver their presentation with compassion, clarity, and confidence.

Does this perhaps sound like you? Are you on this progression toward becoming an enlightened media expert? Are you feeling the need for a larger platform, a wider audience, and a truly powerful response to what you have developed?

After all the preparation work that you've done, if you don't learn how to succeed at the enlightened celebrity level that brings your product or service, wisdom, or guidance into the public eye, then really—what's the point?

I'd like to offer you, in this book, a concise pathway that you can follow in order to bring your expertise forward and offer it publicly to the world. This path is definitely not for everybody because it does require the courage to break through the usual human inhibitions and challenges and dare to be great and exposed and a champion of your cause.

Follow Your Inner Voice

Becoming a noted expert requires a definite calling to come forward and be seen as a bright light in your field—but I've found that, usually, the call to become a recognized expert starts quietly; it's more of a persistent whisper, a growing realization deep within you that you have something of importance to contribute.

At some point you realize that your life thus far has been a laboratory of sorts. You have cultivated a level of experience and insight and know-how that is not yet available to the average person. And you realize that there exists what I call a "hole of experience" that needs filling—and you're someone who can fill it to society's benefit.

Let's take a moment and look at the word "enlightened." **According to the dictionary, enlightened means to give knowledge or understanding to someone; it is to explain something.** In a spiritual context, it means to hold more light. Light, is the purest form of transformational energy. When light is added to any circumstance, experience or teaching, truth is revealed. Light is generated within an individual by an understanding on a conscious level of love and compassion.

When you choose to become an enlightened media expert, you're choosing to be perceived as a visible leader who can help guide others to a higher expression of their own potential using your awakened consciousness to express principals of love and compassion that will help end their suffering.

For example, someone with a personal passion builds a school in a town in Uganda and uplifts the lives of children who live there—and then realizes that he or she also has the potential to become an enlightened expert on poverty and education and uses the media to spark a much larger flow of attention and resources in this much-needed direction.

In order to succeed, every cause, charity, movement, and idea needs a face, a story, and a voice—because human beings are hard-wired to respond with compassion and action to a compelling story with a deep personal consequence.

In this book, I'll guide you through the various steps that I've identified and evolved during my own journey into expertise, steps that for most people are definitely required in order to become recognized as a enlightened celebrity expert. The celebrity is the catalyst, which sparks the attention required to make a shift in circumstances. There are many dark places on this planet which require an infusion of light to transform a long-standing history of hatred, suffering, and pain. It is not personal celebrity that we are discussing in this book; often it is a cause or a movement that is in the spotlight. However, this being said, every powerful vision for transformation requires a leader who is a spokesperson with a voice of clarity to express the ideas and philosophies required for sustainable change.

No matter what your expertise might be, you can learn to harness and fine-tune your story (either one you have lived yourself or one you have witnessed) and then become the visionary mega-phone initiating the actions needed to advance and fulfill your story and your passion.

By learning to master such step-by-step pragmatics of creating content for publicity and media to establish your leverage and authority as an enlightened celebrity expert, you will readily develop the operational strengths you need in order to step forward into the public eye and inspire the world to act in harmony with your cause.

If this opportunity rings bells with you, then turn the page and we'll begin the realistic process that will advance your vision into manifestation on the world stage.

Part 1:
Mindset

"Look at your life and

share your story,

it may be a map for someone

who is lost."

Kristen White

Understanding the Mindset of Voice

Let's begin by defining voice. The voice can be described on many levels: your physiology, which creates a unique sound that comes through your mouth; your emotions, which connect to your brain that are then interpreted into a language of expression carried through your voice; and your intellect, which connects through your mind, creating a persuasive message that is conveyed through your voice to attract support for your ideas, concepts, and beliefs.

Your voice is the most powerful tool for personal expression that exists. When we can harness our emotions, connect to our inner wisdom, and utilize our intellect to all flow at one time to the channel of our voice, we can create a movement that magnetizes the masses.

The Physiology of Your Voice

Your voice is unique. Voice is created when air is generated from the lungs and then flows across the vocal cords, which then vibrate creating a unique sound exclusive to you. The uniqueness of your voice is so profound that if you were to call a dear friend you had not spoken to in more than a decade, the chances are strong that they would recognize your voice immediately. Often-times, when we connect to other people our voice is imprinted in an area of memory within the brain that is easily recalled when heard again. Think about this for a minute. Your unique voice automatically imbeds within the mind of individuals with whom you have made a connection.

Your voice is emotional. Your voice is an emotional connection that can be positive or negative. When your emotions are engaged, the tone and vibration of your voice are directly impacted. Often when we have fear it translates through our voice. We become short of breath, which means less oxygen flows over the vocal cords, leading to a thin, wavering, sound that comes from our mouth. Individuals who experience fear and anxiety frequently do not have a strong, deep, rich voice. Likewise, when we are angry, our adrenaline is pumping, and we have a force of air which flows past our vocal cords, creating a scream that can be characterized as piercing, aggressive, or sarcastic dependent on the internal interpretation of the listener.

You may notice that some people speak with increased volume as a way to attract additional attention. However, a loud voice without a properly energized message is often perceived as an irritation. This creates the reverse response than is intended. Ironically, a soft voice, or an average voice conveying thought sparked from an inner source of wisdom and inspiration is often perceived as a crystal-clear and enduring message by the listener. This leads to a conversation about persuasion.

Your voice is persuasive. When we are speaking, even at a very early age, we learn to create language that attracts the attention of our parents. This language is a powerful tool to help us meet our personal needs. To let our caregivers know we are hungry, tired, cold, excited, or happy. If you are reading this book and you have children, you instantly can compare your response rate to a screaming unhappy child, a polite, well-mannered child, and a loving, inspired child. Who do you pay attention to? Which tone activates your desire to cooperate?

Persuasion is an important tool in any business or professional organization. In the field of law, it is the power of voice that is used to argue on behalf of a case. In the field of marketing, it is the power of voice that is used within a clever construct to attract attention to a brand. In the field of literature, it is voice used within the context of metaphor that inspires the emotions of the reader. And in the field of charity, it is the voice of an individual experiencing a struggle and sharing that story that attracts donations. Ultimately, voice is the key to persuasion that leads to action.

The Importance of Voice

To understand the importance of voice, we need to look at what happens to people who are unable to speak or people who speak but are unable to be heard.

Allow me to take a moment to share a personal story with you. Soon after the birth of her first and only child, my great-grandmother never spoke again. Her husband, her daughter, and her family could not get her to respond to any of their conversations. The doctors examined her and there was nothing wrong with her vocal cords or with her physical health. She simply stopped talking to everyone.

After failing to speak for several months, she was placed in an institution where she lived for almost 40 years. One night, toward the end of her life, a nurse reports that she requested her to turn out the light as she left the room. These are the only words she said for the rest of her life. It is a mystery within our family. No one knows what happened to her. Now, I can only speculate, that a tragedy or a personal attack occurred which led to a personal belief so strong that is was deemed "unspeakable."

Even today, there are many victims of crime, rape, and incest who feel that it is unacceptable to use their voice to speak of their attacks. Likewise, there are many people, perhaps you are one of them, who have experienced a traumatic or painful situation which has generated a deep sense of shame. Shame keeps you silent. However, shame also creates an illusion that you are the only person on this planet who has had this experience.

The truth is that every experience we have personally is a shared experience with humanity. Thousands of people on the planet have had experiences just like you and they are keeping their voices silent as well. If you are someone, who is holding a painful truth, your voice is the most powerful healing tool you possess. When you express your truth, you heal yourself. Then when you express and interpret your truth as a lesson to help others, you inspire them to be healed.

Right now, we are talking through the lens of an individual. But this perspective is also powerful for a group, an organization, or even a large corporation.

The first-person story told through the singular voice of an individual is the most engaging, magnetic, and powerful persuasive tool in the world.

Voice is the essence of story. Stories, especially real-life stories, create an emotional connection that embeds within our memory and, depending on the depth of the trigger, can linger for a lifetime within the individual who makes the connection to the content of the message.

Today, we see a surge in the popularity of reality TV. This is because these are real people behaving in a way that creates a distraction in the culture through the use of the spectacle. Interestingly, if you are a follower of reality TV and you attempt to recall what occurred with the characters six months ago, the odds are the drama has evaporated from your awareness. However, if you attended a dinner and listened to a speaker share her story about her family growing up in a Cambodian concentration camp, you would probably remember her vivid emotional details even a few years into the future. Why? Because one messenger activates our emotions and compassion while another activates our ego. In reality TV we are engaged with our egos, comparing ourselves to the outrageous behavior and story line and ultimately reaching conclusions about our own lives. This is why our memory is not activated, because the thoughts and interpretations are still about ourselves and part of our own ongoing inner dialogue about how we define ourselves in comparison to the rest of the world.

Voice is about engagement. In order for a message to attract us to the point of action,we must be engaged on a level much deeper than the constant inner chatter that conveys us daily messages about ourselves, our inner circle, and our identity. The truth is that most people are consumed with thoughts about themselves.

To break through, a voice needs to make a connection and then create a spark of attention. These penetrating messages come through people who ignite emotions within us that are impossible to ignore.

Voice activates. Once engaged and activated, these emotions will percolate and create a pressure that grows until there is an opportunity or direction on how the individual can release them through action. This expression can be to make a purchase, make a donation, join a community of like-minded individuals, or research more about the thoughts conveyed by the messenger. This will give them the information needed to take one of the actions mentioned previously.

The most effective messengers on the planet instigate action. This has historically been the case. In the world we live in, to be an effective messenger means we need to connect our emotionally engaging message to a specific set of actions for the inspired listener to follow.

History has shown us repeatedly that one person can start a revolution. One voice can inspire a movement. One book can create a trend. Is your voice the one the world is poised to hear?

Not too long ago, many cultures would punish their outspoken members by cutting their tongues from their mouths. This was the way to immediately silence a rebellion. Fortunately, we no longer live in a time where we are punished for speaking what is on our minds. Now, we find ourselves in environment with an abundance of voices with both competing and conflicting messages. This creates a different dynamic altogether.

The universe hears your voice and responds, quite specifically, to the instructions that pass through your lips. Your voice has the power to transform, inspire, and heal others. The words that you choose should be selected with the intent of love, healing, and compassion for others and for yourself. These are the words that have the most magnetic force and penetrating healing energy. If you desire to be an enlightened expert, you must have a conscious awareness of self love. If you don't love and care for yourself deeply, you will burn out. You will not have the internal stamina to go the distance. Self love and spiritual connection are where the fuel comes from to launch your rocket to success. God is your PR manager, but only if you request support.

Voice Activation
Exercise #1

Meditation and Mindfulness

I believe when we have a desire to be a recognized thought leader, it is important to anchor ourselves daily in a private practice of meditation and mindfulness. This allows you to stay grounded and to access clarity and divine inspiration.

I like to use a daily spiritual practice I learned from Wayne Dyer. It's called Joppa. Every morning, when I wake up, I go to my chair and use this simple practice. The practice is stating the word "aum" or "ohmn" several times over a period of twenty minutes. This word has incredible vibrational power. You can feel the energy rolling through your body as you speak it aloud.

*"The people that make a durable difference in the world are not the people who have mastered many things, but **who have been mastered by one great thing.**"*

John Piper

Expressing Your Authentic Voice

There are three key voice expressions that a person needs in order to emerge as an enlightened celebrity. Let's begin by taking a look at these qualities so you can see right away where you currently stand in your evolution toward bringing something of importance out into the world.

The first expression is internal. This is what I call being Recruited by the Universe. Life circumstances, expected or unexpected, triggered an immediate deep and personal voyage down a path of no return. As you moved through this significant, insightful inner or outer journey, your own life was a hands-on laboratory. The impact of this real life challenge may relate to your health, your business, your relationships, or your finances. During this journey you have had to look deeply into these visceral life altering experiences, and through this reflection you have step-by-step become knowledge-able and wise beyond what you could achieve inside a traditional academic learning environment.

In other words, you've been educated in the school of life, and you've had first-hand encounters that led to significant new real-izations regarding the world's problems and possible solutions to those problems.

The second expression is that you have been able to trans-late those experiences into a formal level of expertise, into a tangible process or tool or insight that enables you to support, educate, and benefit others with their own journey. If you remain deep within your wound, you will never truly inspire any reaction to your plight other than pity and compassion for yourself. If your message is merely a harrowing personal story, it is limited to pro-voking emotional, judgemental, or critical reactions, depending on the audience. Our personal pain has the ability to blind us to the reality that we have become identified with our experience in a way that has robbed us of perspective. In fact, we can move through an experience and arrive at a vantage point where we are able to extract wisdom. This is the transformational power that in turn flows through us as an expert. We become the guide for others.

Imagine being on a journey with a guide who was breaking down into tears at every bend in the road because the landscape reminded them of a painful past. How long would you follow his individual?

You will need to arrive at that point in your life where your journey and insights are no longer just about you and what you've discovered in your personal experience, and you are ready to move your realizations and tools forward into public presentation and application where they can serve others.

The third expression is what I call the "It Factor." Every noted expert of all genres has developed this particular quality of confidence and power to the point where people say "she just has it; that's all I can say." It is the subtle and in many ways mysterious "charisma factor" that enables you to shine in front of the camera and attract people to you and your cause.

There is another element to the "it" factor. It's called *Vibrational Credibility*. We all send out waves of energy all of the time. This is our energetic signature. People read our energy when we walk into a room, when we speak, and in the way we move our bodies. Then they decide if we are an individual they like, trust, and want to listen to. Vibrational Credibility occurs when you are in complete alignment with your message. It's when your non-verbal energetic vibes match your verbal message, and you become saturated with your authentic truth. A deeply saturated persona is irresistible to the public.

To be an enlightened media expert you must learn how to detach from your own pain and expand your own personal experience into a larger universal and impersonal perspective where you resonate, connect, inspire, and teach all those people in the world who are eagerly and often urgently seeking what you've found.

The Authentic Expert Mindset

Let's look a bit deeper into that nature of this mysterious attractive radiance that comes flowing out when you learn to tap your true authentic nature and become magnetic in a way that naturally attracts people that you can realistically help.

Many people have their remarkable personal journey, make important discoveries about life, then work to translate those realizations into a transferable teaching or beneficial tool or product, but fail to move through the process of tapping their inner charismatic power.

We will be talking throughout this book about how you can consciously become more powerful, integrated, confident, and aware. You can definitely learn how to tap your deeper core of being and access at will the full "It Factor." This will enable you to more successfully attract others to your cause.

You have probably met a number of people who seem to effortlessly capture the hearts, minds, and support of those they want to teach and serve. Whether you realize it or not, you also have the charisma seed-energy naturally implanted within you.

What is necessary is to learn how to cultivate your charisma. Focus your inner power and on demand send it out into the world. This is how you become perceived rightly as an enlightened celebrity in your area of expertise.

When you reach the turning point where you do truly have something of significance to offer to the world, you are finally in optimum position to rise up and broadcast your story and message loud and clear.

Much of the discussion in this book will focus on how to shift from receive mode, where you gain life experience, into broadcast mode, where you spread your unique vision in the world.

Set Your Inner Radiance Free

As we'll see, in order to broadcast loud and clear, you must first become centered and clear within your own heart, mind, and soul. And if you are like most people, you'll probably need to spend time preparing for your upcoming broadcasts by doing specific inner work that enables you to heal old emotional wounds, reformat distorted attitudes, and become more whole inside yourself.

Like most people, you have perhaps gone through life judging and editing and inhibiting your true nature and expressive power. You've all been through far too many bad experiences that seriously beat you down. As a result, you've denied and repressed much of your primal power. Now is the time to move through the process of reclaiming positive attributes that you might have edited out of your life.

Maybe your voice was criticized, and so you shut down that aspect of your expressive power. Maybe your ideas were laughed at during a sensitive period of your upbringing, so you are afraid to speak your mind openly and clearly. Maybe you thought you looked funny or were too fat or skinny or whatever…

A major step in reclaiming your charismatic power will be to begin to honestly observe your reflex habits of contracting, pulling back, and shutting down, right when it's time to expand, move forward, and turn on.

If you were wounded and hurt, and in reaction you diminished yourself in many different ways, even though it's now time to take your unique experience and teach others, you're perhaps not fully able to be present and powerful because you've left valuable pieces of yourself behind, scattered across the battle-land of your lifetime.

As a result, you might not have much available bandwidth. You aren't broadcasting a strong signal. In essence, your natural charge has been diminished. It's not that you don't have the capacity or potential to be charismatic—it's always available inside you—but to reactivate your expressive power you need to re-learn how to nurture and build your inner charge so you can then release it in attractive public expressions.

To broadcast a wide-reaching signal, you need a certain amount of sustainable energy. This inner mega wattage is what seekers pick up on and tune into. Much like a radio, you need the band-width to receive the signal and to establish a clear channel to transmit your message.

The Two Sides Of Charisma

In order to access and express a strong magnetic presence that will propel your cause or product to the media forefront, there are two quite different dimensions you'll want to explore.

First of all, you must take time to heal old emotional wounds, let go of negative attitudes that don't serve you, and reconnect with aspects of your personality that you edited out of service in years gone by. Charisma requires sustainable reserves of energy, and when you are still carrying around unresolved emotional distortions and contractions, much of your energy is unavailable for positive expression.

We'll be offering steady suggestions for how to look back on past experience and nurture ongoing inner healing and integration. But along with healing the past, it is also very important to nurture all the unexpected present-moment "just emerging" dimensions of your personality.

Life is a journey of continual self-discovery if you take time to pause and consciously open up to receive the newness and insight appearing in each emerging moment. Significant inner growth happens quite naturally, but only when you learn to temporarily quiet the otherwise constant chatter of your thinking-plotting-worrying ego mind, and listen to your quiet inner voice of wisdom and realization.

When you regularly take time for this inflow of creative vision from your deeper self, you do just that—you deepen your presence. And from that deep space you can transmit your expertise. Tapping your true inner voice is what activates the "It Factor" as you express your message.

This combination of inner healing and inner discovery is essential for finessing your media presence. Almost certainly there are youthful strengths and qualities that you have put aside in the past that are valid and needed today. Also, there are just-emerging qualities of your full potential that you have yet to experience and welcome inside yourself.

You don't want to be a one-hit wonder or one-time superstar in your profession or business, where you're not able to sustain your resonant presence under the pressure of the spotlight. The level of focused attention that you put into your personal healing and inner growth will be directly related to your ability to transmit your signal strongly and clearly for the duration.

Honoring Your Experience

When you step forward as an expert, in essence you are purpose-fully offering to hold creative nurturing space for somebody else's growth and success. That's the role of teacher and coach. The quality of space you can hold for others is seriously limited if you're still cluttered inside with your own negative stuff.

Here is an example: A previous coaching client had been working in Washington, D.C. for a number of years as a powerful negotiat-ing attorney. She had an office on Pennsylvania Avenue and was very successful. She realized that the whole lifestyle was making her sick and unhappy, so she left her career and moved to Hawaii. She went off the grid for fifteen years and then came to a point in her life where she felt ready to return to more active participation in her business.

But when I first started working with her on her re-entry, she was determined to block out any aspect of herself from her previous work and life. She wanted to reject her past and be brand new. From the perspective of the enlightened celebrity, that was not a wise stance. She was basically saying, "Oh, that whole big aspect of my life didn't happen; I deny it all."

By discrediting and rejecting such a large part of her life journey, she was losing out on all the gifts, the knowledge, the understanding, and wisdom of having journeyed through her personal experience, regardless of what it was. Only by accepting and integrating her past experience was she able to begin to tap into her true power.

Of course, you don't want to repeat negative experiences caused by old emotional contractions and distorted attitudes you inherited from your upbringing. But remember this: You don't need to identify with and repeat past experience in order to reap its valuable lessons.

I feel certain that every action you've taken in your life, every experience that you've had, every encounter, every conversation, every relationship has offered a contributing resonance to what you are doing today. It's important to put away self-judgment, and honor who you are and value all the experiences you've had.

So in this book I'm encouraging you to go ahead and love yourselves just as you are, and stop denying anything in your past. You surely don't need to continue with behaviors and attitudes that you can see don't serve your present life, but it's a major mistake to stand in rejection of your past experience.

After all, those past experiences remain part of who you are—that's just reality. Along with your genes, your past is the given in your life. And if you choose to love all of who you are wholeheartedly, then you can consciously learn from your past, and in the process evolve into new attitudes and behaviors that serve you better.

Imagine as an example that you're a handyman and you throw away all your tools because you don't like something you built. Such rash behavior won't serve you at all. Instead, you can let go of the design that led to an unsuccessful building experience, and perhaps get some new tools, but keep all the tools that will serve you with new projects.

How can you actively reconcile with those pieces of yourself that you've earlier rejected and pushed away from your present sense of who you are? How can you regain a sense of wholeness that in turn enables you to resonate with confidence, authenticity, and power?

This self-reconciliation process may take a number of different forms, always including inner work where you accept and forgive and thus de-charge emotions and memories that you at one point rejected and pushed away.

So many people who want to be enlightened experts fail because they remain chronically inhibited by self-criticism and negative attitudes, continuously punishing themselves by habitually putting themselves down rather than boosting themselves up.

The enlightened expert mindset involves a conscious choice and action to observe, and accept, and then put aside such self-defeating thoughts and habits and actively reinforce self-healing and self-strengthening thoughts.

Left unaddressed, your clients, readers, and followers will sense your turmoil. This is the Vibrational Credibility factor previously mentioned. Your message will have the reverse effect of what you desire Instead of attracting a loyal following, you will repel people, even if the words appear to be correct. Or worse yet, you will attract the "mirror effect" where your clients and readers have the same level of personal chaos and your business will be burdened with drama and conflict.

Voice Activation
Exercise #2

Authentic Self Discovery

Let's explore a simple process you can learn to integrate into your daily routine, where you tune into some memory of your past that's an emotional charge to you or something that you try to deny as part of your past experience. Here's how to activate an inner process that will defuse and integrate this memory into your full authentic self.

See what comes to mind when you relax a few moments, and then say the following statements to yourself and observe with no judgment what the statements stimulate within you. This inner experience will help generate the integration of your denied past into your emerging present.

Moving through this process often will help you feel whole, which is what being authentic and charismatic is all about. As with all the guided experiences offered in this book, begin by just relaxing.

Tune into your breathing ... make no effort to breathe ... allow your inhales to flow in smoothly and deeply and your exhales to flow out effortlessly and completely ...

Notice that right in the middle of your present breathing experience, there are almost always emotions *... feelings from old experiences living right under the surface of your awareness, waiting to be acknowledged so they can rise up and flow out of you and be gone ... as you stay aware of your breathing, again with no judgment, expand your awareness to welcome whatever emotions you find inside you right now ...*

After a few breaths, see what happens inside you when you say to yourself silently the following words, *"I'm aware of this feeling inside me ..."*

And now, while breathing into your evolving experience, go ahead and say this statement to yourself, on your next exhale, *"This feeling is part of who I am, so I accept it ..."*

And as the final step in this process (which you can move through often during the next weeks to stimulate growth) say, *"I accept the past experiences this feeling comes from, and I now move on with my present life ..."*

Pause and Experience

"You may say I'm a dreamer,

but I'm not the only one.

I hope someday you'll join us.

*And **the world will live as one.**"*

John Lennon

Unleashing the Power of Your Unique Voice

Celebrities are called stars because they shine whenever they step into the spotlight, tapping at will into their inner brilliance and radiating natural warmth that attracts those around them. But for most people, it can seem daunting, perhaps almost impossible, to muster a radiant public presence upon demand.

Many of you don't even like the word celebrity as it relates to yourself or your role as the leader of your business. You prefer to be "low key" and let other's shine. This is deeply ingrained in many cultural traditions. In the United Kingdom, the phrase, "Don't be the tallest poppy," was used to keep children small and silent. We learn at a young age to be seen and not heard. We've been sent messages from our governments, churches, and institutions that it's wrong to stand out. To be noticed is to be revolutionary. However, I believe, to be noticed is to be evolutionary.

No longer is it up to a few elected leaders, billion-dollar corporate platforms, and eroding religions to tell us the few approved solutions to our suffering. Or worse yet, that pain within our communities and our families is inevitable and therefore should be tolerated and accepted.

Learning to turn on your radiance doesn't have to be personally painful nor draining—it can in fact become a pleasure you look forward to in eager anticipation.

Your voice has the power to contribute a solution to a situation that is ready for a shift, but it needs someone, like you, to take notice and then speak up in a loud clear voice with visible call to action.

---

*What do you have to **say**?*
*Who do you want to **be**?*
*What are you destined to **do**?*

Reconciliation, Authenticity, and Breakthrough

You do have a natural brilliance that you can access when needed. You can unleash your unique presence if you move through three phases of preparation and presentation: *Reconciliation, Authenticity, and Breakthrough.*

Reconciliation. The first step, as we have been discovering, is all about reconciling the energetic emotional parts of yourself that you are habitually excluding from expression. Going through this reconciliation process will benefit all aspects of your life—from your love life, to your creative power, to your public persona—and we'll expand upon this process in this chapter.

Authenticity. The second and often parallel step is to begin to move into the space that I call embracing the true authentic nature of yourself. You, like everybody else, possess incredible dimensions just waiting to be tapped and brought to the fore. All you need to do is have this conversation with yourself: "What is there about me that's a diamond in the rough? What's truly magnificent about me if I dare to unleash it?"

We tend to go around fixated on what's wrong with us, rather than what's right with us. From early childhood, most of us were programmed to fixate on negatives, on worrying and self-judgment.

I'm going to encourage you to put that negative self-talk aside—permanently—and instead start talking to yourself in non-stop positive mode, with specific words that will work wonders for your upcoming self-expression.

Breakthrough. Then we come to the third part of the Voice process—reaching the point where you break through courageously and give yourself permission to actually go out and share what you have to offer. This is the phase where you step up and speak out, adding the required but often-missing voice to your inner vision. First—and people so often overlook this step—you must take time and summon the self-love to forgive and embrace the parts of yourself that you've been denying and suppressing as not valid. As this chapter highlights, you need to overcome and transform those haunting aspects that you were led to believe were embarrassing or fraudulent.

Once you integrate your buried power into your present posture in the world, you can move into the space where you begin to naturally beam and radiate. That's how you get that charismatic star quality that shines out into the world and attracts people to you. You stop all the self-sabotage and just haul off and love yourself just as you are, and in the process unleash remarkable inner power and magnetism.

This process is an essential life journey. There is a piece of it that is healing old emotional wounds and negative attitudes and an equal part of the journey that's revelatory, that's out and out spiritual. It doesn't really seem to matter which spiritual or reflective tradition you use, as long as there is a deep, sustained, contemplative and meditative quality to the process where you're able to go inside and excavate that undiscovered or buried part of yourself and set it free.

Moving Into Alignment

I think that the most important first step in this process is to work with what I call *the three levels of inner alignment.*

1. **You'll want to start by clarifying (and perhaps modifying) your core level of attitudinal alignment,** your underlying beliefs and assumptions about what life is all about.

2. **Next you examine your present-moment values and sense of purpose and intent** to make sure they are in alignment with your deeper beliefs and commitments.

3. **Then you actively identify and align the actions you want to make in fulfilling your current cause,** acting in resonance with your underlying levels of faith and commitment.

When you are in alignment on those three levels of personal being, you are ready to go into action. You've made sure there isn't conflict within you, and if you found places where you're out of alignment, you've grown to a place where you attain a higher alignment that lets your energy flow from your core outward into the world without resistance.

For example, a lot of people may want to be a successful coach on relationships, and perhaps they've even written seemingly insightful books on relationships, but when they're honest with themselves, they see that there are relationships in their own lives that are still a mess. This happens so often. People are out there with a desire to be a noted relationship expert, but they haven't taken the time and work to take care of their own situation at home.

As long as their inner alignment is out of balance—where they might intellectually understand what they're talking about but haven't integrated their insights into their own lives—they're not going to have a full flow of manifestation energy emerging from their core, and they'll be sending mixed and undercharged signals in their media communications.

So it's important to "save the world" by starting in your own home. Then, when you're congruent and you are in alignment, you will see it reflected back in the amount of people who come into your professional circle and are magnetized by your genuine presence.

If you're already out there trying to hang your shingle as a successful coach, an author, or speaker, but can't seem to find adequate clients, then I encourage you to go back to your alignment issue and focus concerted attention during the next weeks and months in that inner direction.

Too many psychologists, coaches, therapists, and teachers are trying to do their inner work in their office by working in a laboratory of other people. However, it's always problematic to build a sustainable business when inner alignment hasn't yet been achieved.

As perhaps you've noticed, many people put on an outer image of being experts, when what they're really looking for is an expert to help them.

To truly shine, the experience of inner transcendence is required, where you move through and go beyond your own emotional journey, and are then in position to translate your experience into your expertise.

I've found that the most successful people are the ones living within their own meaningful relationship. Their clients come back again and again and also refer clients because the teacher is trustworthy deep down and grounded truthfully in her or his core principles and identity.

Expressing Genuine Presence

There is a lot of external pressure that's always trying to influence your inner sense of who you are. For example, if you are married, there exists a persona or a perceived identity around being a wife or a husband. Similarly, there is a certain identity attached to being a therapist or whatever other profession you're in There is a certain identity or perception to being an athlete, a sales person, an entrepreneur, a doctor, and so forth.

Given all these external pressures, when you haven't developed a deep understanding of yourself beyond role identities, you will not express a full genuine presence. You're constantly living and expressing yourself through other people's perceived identities about who you should be.

This is like being a chameleon, molding oneself to fit external expectations, and that's not going to generate a solid charismatic presence. You won't come across as authentic because you're not yet coming solidly from whom you really are.

In contrast to that, if you are adequately full in your understanding of who you are in your core being, when you connect with your clients you will express this as if you're not even aware of all the societal norms. **You just come from what your truth is. You share very cleanly your experience, your journey. When you are aligned with your inner values, what you have to offer flows outward naturally in a way that is a benefit to the other person.**

Always remember: it's not about you, it's about them. That's the connection point. Your personal stories and teachings must come from a space that is not identified by any outside norms or self-centered expressions.

When I was growing up, my mom used to always say, "The world embraces the individual." When you're not afraid to just be who you really are, to express your unique personality fingerprint, you find yourself in your own zone. There is no competition, and you can speak about anything with unlimited clarity because wisdom and insight just flow through you.

As you practice the art of stepping forward and learn to spontaneously express your authentic truth, you will soon find that there's an unlimited amount of insight available to you; whereas somebody who is working from identities that have been superimposed on them will say the same ten to twenty statements over and over again, speaking from their robotic, automated response system.

The Push To Become Famous

In this process we're exploring together, you'll want to think seriously about what it really means to you to become an enlightened celebrity. A lot of people desire to be a celebrity of some kind because they feel that at some point in their life they were excluded, they were seen as nobody. I've come across this so often. Something happened in their childhood that put them on the sidelines and marginalized them. In the back of their mind there is an unresolved and often unconscious emotional wound, and the defensive stance that silently mumbles, "Well wait until I'm famous someday, wait until I get out there. I'll show you!"

Too often there's a compulsive attraction toward stardom that comes from a space of revenge. It's like, "I received this wound in my life and people treated me as if I didn't matter at all. I've had this painful journey for too long. Now I want to be super successful so I can go back and show those people who put me on

the sidelines that I'm now recognized as important in the world."
That's celebrity revenge, and deep down, it never satisfies.

Of course, there's also the other side of the celebrity role where,
when you attain the spotlight, you feel positively empowered
to serve others and further the higher good. Right now there's
incredible service happening on in the world through high visibility celebrities. They're using that visibility to save children who
are starving, to bring medical help to the impoverished, to help
make the world a better place.

If you feel the calling to become an enlightened media expert,
no matter the size of your niche, it's important to make sure you
consciously transcend any negative reasons you might have for
wanting celebrity status. It's a role that can be one of the highest
paths of service when it's entered into cleanly and positively. If
your true heart intention is to make a contribution and an impact
in the world, then the power will come to you.

If your intention is to attract celebrity attention so that you can
help other people by your visibility and your voice, then almost
surely fulfillment will come to you.

The Enlightened Expert

I know a young mother who ended up spending three years living
in a hospital with her two small children, Matthew and Andrew,
who were both extremely ill. She was a very devoted mother and

she would never leave their side. And while going through this tremendous ordeal, she observed the hospital world and kept extensive notes about her challenging situation, recording everything that she learned while being in the hospital.

She certainly didn't ask for this traumatizing experience, she was just a stay-at-home mom. But she was recruited by the universe to go and live in a hospital for three years, and while moving through this life journey, she became an expert on being inside the hospital culture and caring for a critically ill loved one.

Tragically, in the end both of the boys died. And there she was, assessing what had happened to her, losing her two children. Her grief was overwhelming and she needed to identify meaning in her personal experience. She'd meanwhile had the unique experience of living in a hospital all that time. Where could she find meaning? What was the purpose?

She devoted herself to turning her notes about hospital life into a book, to understand what she'd learned from the experience, and perhaps to share it with others. After all, she'd had this incredible journey and she wanted to somehow contribute. She turned a terrible period of her life into something meaningful by developing her unique expertise and putting it into a book.

Suddenly she found herself an enlightened celebrity expert, with thousands of parents in similar situations seeking her out and being helped, guided, and comforted by her words of wisdom.

She rose to the occasion, even though she was naturally quite shy, and created a foundation for which she was the media spokesperson. People were calling her from all over the world to learn more about what she knew from living in a hospital environment while caring for her critically ill children because it's such a complex culture.

Now, she could have just stayed home and said, "Well, I've lost everything, I'm going to go hide in the closet and just wait for my life to end so I can join my two boys." She could have done that, and there was a point where she wanted to do that. But instead, she said, "I have this expertise. I have this knowledge, this powerful journey—traumatic as it was—and I know it can inspire and benefit others who are caring for critically ill loved ones."

So she wrote her book, and in the process looked deeply to her own spiritual core, and brought herself into a place of wholeness.

Did she then raise her hand and say, "I want to be an expert on this topic?" No, it wasn't that easy; she had to move through the inner process of getting clear of her motives, solidifying her voice and message, and then stepping into the scary limelight and beginning to touch thousands and thousands of lives, on a profound level, by sharing her journey.

She took a life experience, a dawning revelation for her, and chose to commit to something that had significant meaning for her, and in the process she has helped a lot of other people. She heard a voice deep within her saying, "I feel moved somehow to express myself in a unique way. I want to give my unique gift to the world."

It's important to note that this woman was happy to share the information. She wasn't attached to it, she released it freely. And she wasn't expecting other people to give her unconditional love for her sharing. She wasn't saying, "Love me, love me." She just said, "This is my journey, this is what I learned, and this is how I can help you, these are the important steps." And she took the initiative to develop and place her insights and wisdom inside an organized system, inside a framework that educates and teaches others.

Truly, the enlightened expert is more about "what I can give" than "what I can get.'

It's not that these people aren't being compensated or they're not getting some remuneration to their charity or whatever. Everyone deserves due recompense for their offerings. But it's essential to reach a point where you're coming in a heart-centered attitude related to giving versus taking.

The enlightened media expert that I'm talking about wakes up every single morning and asks the question, "How can I serve today? In what way, big or small?" And they're not always serving in a large way. Sometimes they're serving by being a genuinely joyful, cheerful person as they check out at the grocery store.

Next, they ask themselves, "How can I broadcast my expertise and insights to impact the maximum amount of lives."

Whatever the situation, there will usually be a consistent positive energy that flows through their entire being. They're not just "on" when the world's watching. They're steadily anchored in the natural warmth of their expertise.

Law Of The Circle

As you've almost certainly noticed, part of the mysterious dynamic of moving toward personal fulfillment is the fact that the flow of life regularly brings recurrent themes and opportunities into view. Each time a recurrent situation or challenge appears, you have the chance to either ignore it or respond and go deeper into that particular theme in your life.

When you begin to notice these recurrent situations, you begin to respond positively to this cyclical aspect of how your life evolves and rapid advancement is possible.

Of course, each time you go around one of these circles and encounter similar themes and opportunities to grow, time has passed, so what we're really talking about here is a spiral, carrying you further upward into discovering your true potential and power.

Most of us have experienced this cumulative effect in our life—

which is traditionally called the Universal Law of the Circle. Sometimes you move through small circles. For example, you will meet someone one evening, and then two days later you'll see them again, and then the following Friday you'll see them yet again.

Call it synchronicity or chance or whatever, this cyclical aspect of life regularly offers you a new chance to respond and move forward within a particular realm; each time, you have the opportunity to loosen the restrictive leash of that aspect of yourself yet another notch.

Sometimes in life, you'll also encounter a larger circle or spiral. For example, in my own life I worked on the studio lot of 20th Century Fox when I was in grad school. And now 20 years later, I find myself connecting with some of the same people at 20th Century Fox, because I have a movie that I want to produce about a spiritual journey to a Tibetan monastery.

That is an example of a longer circle. Sometimes circles are only 24 hours, 36 hours, 2 weeks, or 2 months. They can be 10 or 50 years or even longer, but I believe that everything in our lives participates in this circular dynamic.

Think about Bill Clinton. He had his picture taken with JFK when he was a page in the White House—and then there was a big circle—and there he was 30 years later, standing at that same White House as the president. That same type of natural recurring pattern manifests in everybody's life.

But here's the problem: If we're constantly snipping and clipping and editing and rejecting and denying certain experiences and aspects of our past, we end up disrupting our natural recurrence-patterns and never complete our spirals. We end up with an absence of wholeness in our being.

This is why the inner work of reclaiming all of our past experiences is valid and so important. Everything in our life follows a pattern, just like the tides naturally have their cycles.

We go through each phase of a day; we move through spring, winter, fall, and summer; and then we repeat that cycle. We go through long cycles of life, death, and rebirth. But we also have many smaller but important cycles inside our life where themes will come up regularly, enabling us to go a step deeper if we're ready and perceptive of opportunity.

What's important, for recognizing and participating in the spiraling advancement of your life, is being aware, being present in the moment, and responding to the gifts and experiences that come to us.

You need to pay attention in your life through a deeper listening, through not being lost in thought and operating on automatic. How many times have you driven to work and hardly noticed what was happening around you? Who knows what opportunity you missed by not being present to receive.

When you pay attention to recurring patterns, you can say to yourself, "Oh, hey I've been here before. I see how I've evolved and grown since the last cycle. What can I now integrate into insight and wisdom? What's the larger flow here in my evolution and awakening?"

Each cycle you move through, if you're conscious while it's happening, offers you a chance to dig deeper into your expertise. You have latent capacities that you are not even aware of. Studies show that most of the time you are only using about 10% of your mind and possibilities. What might you do with the other 90%?

Developing Your Daily Practice

The key to tapping your other 90%, to developing a more radiant inner presence in your work, is your development of ever-higher self-awareness. The deeper you move with your awareness, the fewer your cycles become in learning what you need to learn. Through becoming more aware more of the time, wisdom, and mastery, it becomes much easier to receive and integrate.

For most people, cultivating enlightened expertise does require a daily spiritual practice of some kind, a periodic pausing where you turn your mind's attention inward to your core of being so you can contact your genuine personality fingerprint.

Through this regular pausing, you are connecting and dialoguing with inspired levels of yourself every single day.

You find that you can do this while just sitting quietly or while exercising, going on a bike ride, taking a shower, walking in nature, or sitting by a beach. This becomes part of your practice. You can also do a daily meditation, or go to church, or recite mantras.

But please, on a regular basis, take some time each day to put your to-do-list aside and actually listen to what's coming up for you. That's the ultimate source of all of your expertise. It comes from your own inner dialogues and your higher inner awareness.

Voice Activation
Exercise #3

Inner Awareness

Here's a basic practice that I teach in my workbook and video training programs that will take you deeply into experiencing the basic daily process we're talking about.

Give yourself five to ten minutes to retreat from active engagement with the world. Just sit quietly, stretch if you want to, and give yourself permission to relax and feel good without doing anything at all. Just be.

Calm your mind's busy thoughts by focusing on your breathing ... tune into the feeling of the air flowing in and out of your nose ... and now expand your awareness to include the movements in your chest and belly as you breathe ... set your breathing free to come and go as it wants to ... let yourself live fully here and now in this moment, holding your breathing most important.

Expand your awareness another notch as you stay aware of your breathing experience. Begin to feel the emotions in your face ... your throat ... your heart... and your belly.

Breathe into whatever feelings you find without trying to change them. Instead, give them love, respect, and space *to be integrated into your present sense of self … be inclusive, don't exclude anything you discover within you … and say to yourself, "I accept and embrace all my feelings."*

As you stay intimately aware of your breathing experience, notice how it calms down. *You've accepted your feelings, so you can just be … who you are … right now. And now say to yourself, "I am open to receive insight and guidance from my deep source of wisdom."*

Let these words resonate within you *… stay aware of your breaths coming and going … and be open to receive a new experience.*

Pause and Experience

"I am only one, but still I am one.

I cannot do everything,

but still I can do something;

and because I cannot do everything,

I will not refuse to do something

that I can do."

Edward Everett Hale

"One person really can

make a difference.

Each person is the revolution."

Bryant McGill

Trusting Your
Inner Voice

As we're exploring step by step, what you focus on day **in and day out is what you manifest in your life. This is a basic law.** In order to connect with and manifest your unique expertise, your place of no competition, you must manage your focus of attention properly. You must focus every day on finding deep guidance and becoming a conduit for that unlimited creative channel within yourself.

In this light, I see unlimited opportunities for all of us because there are so many things that haven't yet been developed, and there is so much good work to be done, if only we take time to listen to our inner guiding voice that knows, from the source, what our purpose is in life.

You're not going to get that essential guidance and inspiration from any outside source. You must focus 180 degrees away from the outside toward your genuine inner presence and power.

Steve Jobs said that one of the primary sources of his success was his intuition. Every advance in design that he made, he said, came through his intuition. And that's what being a enlightened expert is all about:

1. Contacting your intuitive sense

2. Listening to it, and then

3. Acting on it

As you regularly return to that quiet space of introspection, your awaiting inner wisdom will help you break through and become the unique expert that you aspire toward, and there's no competition in that space.

Too many people get stuck in what I call "the getting-ready corral." It's very crowded there, and everybody seems to be doing the same thing: getting ready. They haven't found and expressed their uniqueness yet, so they're worried that "he is more attractive," "she is more articulate," "she's a better dresser," or "he has a better program." But all those kinds of factors are external and get in the way of being effective.

To break free from the crowd, just go on an "outer stimulus fast" every day for fifteen or twenty minutes. Put aside every outside thing or thought and focus deeply within to your own creative incubator. Forget what anybody else is doing or thinkingand discover your special offering for the world.

Focus on where you are spontaneously true, unique, radiant, and brilliant. You'll find that the more individual you are, the more you're able to serve and the higher and brighter your star is going to shine.

Listening to Inner Guidance

It's universally agreed that we all have an inner voice of wisdom, if only we can somehow manage to pause and tap into it. Call it God's guidance, call it intuitive insight, call it what you will, but human beings have this inherent capacity to tune in to an inner sense of spontaneous knowing that resonates from a universal core of wisdom and inspirational insight.

But if we all possess this inherent trustworthy guidance system, why do most people usually ignore its guidance rather than actively seeking and following it?

Here's at least a partial answer to this important question: All of Western civilization for the last two-hundred years has fixated on an educational system and a philosophy in general that over-emphasizes the deductive, intellectual function of the mind and ignores or even rejects the emotional and intuitive functions of the mind.

From the time kids are five years old, they're taught to fixate and trust their deductive reasoning and to ignore their deeper intuitive and integrative feelings, hunches, and insights.

Even though cognitive science fully recognizes that the brain has the power to perform instant integrative functions that generate a trustworthy sense of knowing what's right, what to do, how to respond, and what to feel in a situation, children are taught to be downright suspicious of this function and to ignore their inner voice of wisdom.

This fear-based element makes people afraid to trust and listen to and behave in accordance with any inner notions that come to them beyond the limits of their reasoning intellect.

We have progressively pushed our emotional intelligence brashly away as insignificant and untrustworthy. Therefore, we tend to ignore our gut feelings, our subtle hunches, our emotional and physiological reactions that would otherwise guide us in our decisions and actions.

Hopefully we're entering a period of history where we start to teach our children (and ourselves) how to develop emotional intelligence, how to respect and tap intuitive insight, how to turn on the powerful wise integrative function of the mind. In this spirit we can better listen to our deeper voice and trust it to guide us toward actions that help us fulfill our higher calling.

We must begin with ourselves, right now. We must discipline our daily lives so that we periodically stop and tune into our bodies, our feelings, our intuitive voice of wisdom—and determine for ourselves through observation and experience if our inner voice is trustworthy.

Don't take my word for it—look to yourself daily over the next few weeks, and determine the best way to tap into and evaluate this universal and perhaps essential internal guidance system.

I suspect that soon you'll be able to identify when an idea that pops into your mind is nothing more than a reactionary manipulatory ego response coming from shallow levels of your everyday thinking mind, versus a deeper core insight, realization, or creation emerging from your higher self.

You'll discover in this process that when you seek it, you possess a particular sense of inner knowing that lets you discern correctly if the interior voice within your mind is authentic, trustworthy, and worth responding to with action.

Discernment at this level is directly correlated with the consistency of your introspective practice. If you take five to ten minutes every day and conscientiously practice introspective reflection and inner calm, you find that you can tune into more subtle quiet qualities of consciousness where your inner voice is extremely clean and clear.

Confirmations and Nudges

Your career as a niche expert is constantly in flux. Every day you have a number of choices to make, each of which can determine the direction of your career. With each decision, if you're in tune with your inner voice, you'll have an edge in making the best decision. How can you maximize your connection with your career's intuitive guiding voice?

When I do my intuitive work I always move through a short preparatory routine or practice to begin my reflection time, similar to what I guided you through at the end of the previous chapter. I also end with a short ritual to close the insight session.

Working from within this established meditative framework, I can relax and feel confident that whatever happens in the middle of the session is clear and true, clean and authentic, because it is emerging directly from a clear, honest, receptive mindset that I've established and trust as a primary link to my deeper intuitive voice.

The voice of insight and guidance comes to each of us in our own unique way, and often quite spontaneously in the most curious situations and places. Sometimes we receive insight through symbols or metaphors, or through a dream or daydream that resonates with special meaning to us. Sometimes we'll hear a song on the radio that wakes up an idea or decision.

Insight and growth will come to you through a lot of different channels; it's not necessarily an actual inner voice or thought. Sometimes is comes quite directly as a definite inner knowing. People regularly report suddenly flashing in their mind with a complete answer to a major challenge or a perfect design for a new product. At other times, the inner voice speaks more as a gut feeling or sudden knowing quite outside normal rational thought processes.

But in any case, when the authentic voice comes, it has a clarity to it that you will recognize on a soul level immediately. Curiously, there's usually no emotion attached to it; it just drops into your awareness as a pure gift, which is what it is.

If you ignore it, it will probably come back often. I noticed this with myself when I wrote my first book. I was meditating and doing daily spiritual practice at the time, and my intuition kept quietly suggesting that I was going to write a book. At first I reacted with, "No, I'm absolutely not."

This happened over and over, and then one day I went to a little bookstore where a friend of mine was working and she said, "Oh, this just came in, it must be for you." It was pamphlet for a writer's workshop, and suddenly I realized that yes, I am going to write a book. At that moment I heard my inner voice being reinforced in the external world by a physical sign.

That's what I would call an expression of the Universal Law of the Circle where you regularly continue to get signs, confirmations, nudges, and messages at an increasingly clear level until you're ready to listen to your voice and act on that guidance.

Life is about the impact

you make

with the message

you share.

Find your voice...

Voice Activation
Exercise #4

Listening for Your Inner Voice

In the last chapter, I shared with you a core process for pausing, quieting your mind, and tuning into your feelings. Let's take that process another step now so that you create regular time and inner space for insights to come into your mind where your inner voice can speak directly to you in whatever way it naturally does.

Here's the quiet-mind process expanded into an insight session that will help you on a daily basis to make the right decisions for your career, and life in general.

My challenge to you is to set aside five to ten minutes each day to move through this important process so that you stay directly in touch with your deeper wisdom and guidance. Discipline of this kind will work wonders for you.

By the way, for guided sessions such as this, you'll find that

listening to an audio version is very powerful for going deep. *You can record your own voice reading the following guided session or go to my website for audio guidance.*

Set aside five to ten minutes where you can withdraw from active engagement with the busy world. *Get comfortable sitting somewhere where you won't be disturbed. Go ahead and stretch if you want to, and give yourself permission to do absolutely nothing except relax and feel good.*

Now take a few moments to calm your mind by focusing on your breathing… *tune into the feeling of the air flowing in and out of your nose … and expand to include the movements in your chest and belly as you breathe.*

Relax and set your breathing free to come and go as it wants to … *be fully here and now in this unique emerging moment with your breathing held continually as the most important happening.*

Now expand your awareness another step as you stay aware of your breathing experience, *and at the same time experience the emotions in your face … in your throat … in your heart … and your belly.*

With each new inhale and exhale, continue to breathe into whatever feelings you find without trying to change them at all. *Instead, give them love, respect, and space to be integrated into your present sense of self … be inclusive, don't exclude any-*

thing you discover within you … and say to yourself, "I accept and embrace all my feelings."

As you stay intimately aware of your breathing experience, notice how it calms down. *You have accepted all your feelings, so you can just be who you are right now.*

You're now in perfect inner condition to open up to whatever insights come to you, in whatever format. *Just say the core "insight boost" statement to yourself, "I am open to receive …"*

Hold your focus on these words deep within you, as you also stay fully tuned-in to your breathing experience *because often insights come flowing in on your breathing.*

You might also say something like, *"I am listening to my inner voice."*

As you continue breathing, you can stay in this receptive mode for as long as you enjoy it. *Sometimes nothing will come to you in the moment, but you've opened up and planted the insight seed, so that perhaps an hour later, when you're doing something entirely different, insights will flash into your mind.*

To end your session, begin to explore different closing statements you can make. *Here's one you can use and feel free to explore your own ending ritual to do each time you move through this insight process: "I am thankful for my inner voice of wisdom. I now end my insight session."*

Pause and Experience

"Our personal journey is always

through our wound.

We analyze it, deny it, project it,

and consequently repeat it.

Freedom comes when we face it."

Kristen White

Part 2:
Message

"Each choice we make causes a

ripple effect *in our lives.*

When things happen to us,

it is the reaction we choose that can

create the difference between the sorrows

of our past and the joy of our future."

Chelle Thompson

Mastering Your
Message to Yourself

Most of the time we are busy in our minds talking to ourselves, commenting on everything we encounter, quickly categorizing and judging people we meet, thinking and worrying about problems in the future, or ruminating about upsetting events in our past.

Sometimes we think uplifting thoughts that boost our mood, but all too often we're locked into chronic, anxious, judgmental self-talk that clouds our thinking with upset emotions that in turn generate stress and damage both our health and performance.

Don't be a victim of your "monkey mind'. All these convoluted conversations go on chronically and even compulsively inside our minds, and most people fall victim of their own thoughts. I spent time studying with a Tibetan monk who talks about the monkey mind—that constant chatterbox voice inside us that is, in scientific fact, a root cause of far too much of our stressed-out, worried, confused mindset.

Yes, we are constantly bothered and even almost possessed by our monkey-mind chatter. However, deeper inside us, beyond the monkey mind, we do have the mental capacity to manage our thoughts. For instance, as cognitive therapists recommend, we can learn to say positive thoughts or mantras over and over again to ourselves to override our habitual negative thoughts. And we also have the inherent power to fully quiet the flow of thoughts through our minds.

Self-talk can be toxic to your voice. So much of our self-talk is toxic because it emerges from ingrained attitudes and negative one-liners we inherited very early on in our lives, perhaps even in the womb. No parents are perfect, and every time our mothers sank into a negative, anxious, or contracted emotion while we were in the womb and during early childhood, we absorbed these bad vibes directly.

Even before we knew how to talk and think, we were bombarded with the upset, hostile, frightened, or depressed feelings of those close to us.

And then, as we learned to talk and to think, we were imprinted by the ideas, attitudes, prejudices, assumptions, and worries of our parents and teachers. We learned not only *what* to think but *how* to think from our parents. And as adults, unless we take action, we continue to have these imprinted thoughts running through the back of our minds as ingrained one-liner assumptions and beliefs.

Early-Childhood Imprinting

Consider this, if your mother basically believed about herself that "I'm not good enough," or "life isn't fair," or "I wish I was some-place else," or any one of thousands of other negative one-liners, you took all these disastrous one-liners into your being with your mother's milk.

Cognitive science has proven that your own thoughts will continue to be polluted with your parents" negative self-sabotaging one-liners all of your life unless you become aware of your habitual semi-conscious self-talk, and learn how to silence and replace it.

Chances are very high that you still have language inside of you from your childhood that says you're not good enough, you're stupid, you don't have anything to offer, you don't look as good as you used to, you're fat, you're ugly.

This is all self-judgment, and thinking these thoughts subliminally does nothing but knock you down and depress your mood and reduce your chance of success in life. What you think is what you manifest, and if you're self-sabotaging yourself all the time with your own negative self-talk, you are in essence shooting yourself in the foot on a regular basis.

Toxic self-talk stands as a primary wall between you and your ability to take your positive vision out into the world; therefore you must become more conscious of your self-talk.

You must learn how to focus your attention on positive, uplifting, self-loving thoughts—purposeful self-talk that can override your childhood programming and shift your mental attention in directions that directly reinforce your success.

Anything less than complete self-love is self-sabotage. Except for dispassionate self-observation, every negative thought you think drags you down and is of almost zero value in your life. To see this truth clearly will initiate a natural movement away from negative self-talk.

Negative self-talk broadcasts fear. There's a further problem with negative self-talk in that what you're constantly saying about yourself will begin to broadcast outward. You will project your judgmental fearful attitudes onto the world around you, and this critical attitude will chase people away from you, right when you want to attract them. Dealing with your self-talk habits is crucial in your relationships and career.

How To Stop Negative Self-Talk

You can't really run away from your habitual thoughts, nor can you take a pill to silence the internal internal chatter in your brain. You can try to drink it away, or exercise it away, or trick it, or fool it into leaving you alone, but nothing short of a major shift in how you use your own brain can quiet and transform negative self-talk.

Quieting your mind with cognitive shifting. It's important to explore how the thinking brain works, and then to practice what psychologists call *"cognitive shifting"—the fine art of purposefully quieting your thinking mind by aiming your mind's attention entirely out of thinking mode into present-moment sensation mode.*

The four brain nodes. Your brain has four different modes: *thinking mode, whole-body sensation mode, emotion mode, and intuition-integration mode*. When you use each mode equally during an average day, you're nicely balanced. But when one mode (e.g., thinking) gets out of hand and totally takes over the other three, you suffer, especially when that mode is busy thinking thoughts that make you feel worried, stressed, guilty-ridden, depressed, etc.

How can you regain inner mental balance and regularly quiet the negative self-talk that's dragging you down?

Step 1: Observe

The first step is to begin to pay attention to what you're saying to yourself as you go through the day.

Set a time frame. Maybe one day, maybe three days, maybe a week. Using a note pad or recorder, write down the thoughts you are fixating on during the day, and consciously pay attention to the language that you use.

Our universal disconnect from our inner core is that we are speaking words to ourselves all the time, but we're not paying much attention to what we're saying. In order to get a pulse check of what themes are actually running through your mind over and over again, make note of your habitual thoughts and become self-aware of your self-talk.

Step 2: Positive Self-Talk

If you find that you are continually stating a negative attitude or weakness, actively reverse that statement. For example, perhaps you find yourself over and over again saying to yourself, "I never seem to have time to work out, I just can't seem to discipline myself." Or perhaps you say, "One thing I simply can't do is get in front of a video camera and speak with clarity and power."

Write down these one-liners that reinforce a negative quality of yours. And then reverse the statement. For instance, begin saying to yourself often during the day, "I can find time to work out today, and I have the discipline to do it."

Taking charge of your mind. If you take charge of your mind and think this positive one-liner to yourself often during the next few days you'll almost certainly begin to make time to work out, and discipline yourself to do so.

Why? Because your inner director—that voice speaking to you—is clearly stating your positive intent, and your whole being will naturally respond accordingly.

Positive statements of intent. Let's take the second negative one-liner, "One thing I simply can't do is get in front of a video camera and speak with clarity and power."

Restate this in its opposite incarnation. "I am going to get in front of a video camera today and speak with clarity and power." Once again, you're establishing positive intent. You're putting in your mind the exact words that will maximize your potential for doing what you know you need and want to accomplish.

Many people try to totally cancel or delete negative one-liners they find filling their minds, but this doesn't work. Why? Because you're replacing one negative with another negative: "I'm not going to think that negative thought anymore."

By saying this statement of intent to yourself often during the day, you override the negative thought that was keeping you from acting successfully.

Instead, the higher path, the one that is more permanent, lasting, and sustainable, is to find a more powerful sentence, one that has a higher vibration or frequency, and just substitute it. You take the new one and lay it on top of the old one, and only the new one is visible to the mind's eye.

Creating Bright Living Space

Beyond observing your chatterbox mind and introducing new positive intent statements to override old negative ones, you will also want to periodically move into pure self-nurture where you let go entirely of trying to improve yourself and get ahead in the world, and instead just relax and love yourself as you are.

Self-love is essential. If you're seeking to be more genuine, more authentic and charismatic in your work, spending time regularly in self-love mode is essential.

One of the things that I do regularly for myself, no matter how busy or crazy my life might seem, is I give myself permission every day to take time to entirely empty out my mind. I do this through a visualization practice using symbols because the deeper spirit world communicates this way.

The practice I like is to sit quietly for a moment, tune into my breathing, and then begin to imagine that I'm walking into a room. I just notice what's in the room. Are the windows open or closed? Are there drapes? Is it light or dark? Is there furniture in the room? People? What's going on in the room?

Then, once I've looked around the room with my imagination, I focus on each one of the objects that I notice in the room, one after the other, and I dissolve it. I use a little bit of magic in my imagination and continue to dissolve all the different things in the room so that one by one the room becomes clearer and clearer until it's completely empty.

I then go and open the drapes and blinds and open wide all the windows so that more and more light comes into the empty room, and it becomes brighter and brighter. I then sit on the floor and fully enjoy and embrace the clean, clear, empty brightness.

This is a wonderful way to clear the mind and attain inner calm, peace, and insight. The process allows me to let go of everything in my mind and my life that's causing stress or worry or conflict. I just empty the room of everything, and I let the pure light and air come flowing into my mind.

Often I do this process in the morning to start the day with a clean slate of clarity, peace, brightness, and freshness. The next morning I do the same and clear the space again.

Once you practice this a few times and get good at the process, you'll find that this inner sense of having a clear quiet quality of inner space deep within you is easy to return to and be nurtured by.

Becoming More Receptive

You'll find that as you perform this process and clear a space within you that you bring white light into that space and learn to stay within that white light, and this gives you energy to sustain and express yourself in your work.

Living a creative public teaching and service life is definitely all about wise energy management. The most valuable resource I have as a enlightened expert is my own energetic creativity. And if I don't create space and honor that space for my creativity to flow into and through me, then there's no open real estate for that creative light to come into. I am then seriously cut off from my source and my flow, and my creative expression is dulled and dilated.

If I don't create and maintain that clear inner space, my broadcast frequency is reduced, and my radius, my bandwidth, and everything that I connect to is diminished.

I have found that the clear white light flowing into your inner room of your imagination contains all that you will need for the day. At deeper levels, the core creative and sustaining inflow from the universe empowers everything you do.

It's not important to focus on what you think you want to bring in or what you want to attract. What's more important is staying clear and clean and expansive so that your mind is receptive and recognizes opportunities that flow in upon the wings of that clear, white light.

I know this all sounds rather esoteric, but in practice it's just the opposite. Such visualizations do work. With a little practice you'll find that as you nurture this white clear bright space within yourself, you become more and more strongly guided to what you need to team with in the outside world.

Often what you're seeking will show up right in front of you, you won't have to go looking for it. Your life will become this natural and magnetic dance where, at just the right time, you connect with exactly what and who you need to connect with. By taking five to ten minutes to clear out the energy and let the light come flowing into your life, you will enable yourself to connect immediately with manifestation possibilities. No one knows quite how; it just happens that way more and more when you regularly do the practice I described.

Self-Talk and Negative Beliefs

A person's ongoing, negative self-talk is a constant expression of how they are sabotaging themselves with their underlying beliefs. If you don't believe that you have anything of value to contribute to the world—for instance, if you don't believe it's possible for you to be a bright star bringing forth your contribution—and if you don't believe that you can make an impact on others" lives, then you simply will not.

The question is, if you have self-defeating core beliefs about your own value and potential, how can you shift beyond those beliefs?

Everybody inherits a set of core beliefs from their parents and community, and everybody then advances and develops their adult beliefs—to a certain degree. Most of us do evolve some-what beyond the beliefs we inherit, but is it possible (and is it permitted?) to consciously and actively shed all of your sabotag-ing beliefs, to purposefully change them, throw them away, and replace them with beliefs that serve you better?

The Answer is Yes

My story. When I was in grad school, I really wanted to be a reporter. I wanted to work in TV news as a journalist because I felt that it was a way that I could help other people. I used to call it being a voice for the voiceless, being a champion of people who were struggling in their lives.

My desire was to be a reporter, but everybody I knew said, "You're never going to break into TV; it's not going to happen. You've got blond hair and blue eyes, you're a dime a dozen; that's not what they are looking for these days. They are looking for people who are from different cultures and therefore more interesting."

If I had taken on that negative belief that I was being bombarded with, I would have dropped my deep inner vision and gone off and done something that I believed was easier to achieve and that I was more likely to succeed in but that left me unfulfilled.

But I simply refused to believe I couldn't make it in TV news, even though the odds were seriously against me. Instead, I took a menial job working for no pay as an intern at a TV station in Phoenix. Because I believed I was already a TV journalist deep down, I regularly told myself, "The door is open a crack." Instead of wearing jeans and a T-shirt, every day I dressed up as if I was one of their prime reporters out there who was grabbing the top story every single day.

By holding a positive intent in my mind, my behavior was quite different than if I'd given up my passion. For instance, I made a habit of showing up at the studio when all the other reporters showed up at 8:30 in the morning, just in time for the morning meeting. I would stay through the entire day contributing and participating until the newscast was over, and then I'd go home when the other reporters went home, just as if I were working as a reporter there.

I held onto my positive belief that I was a reporter—I looked like a reporter, I acted like a reporter, I was ready to be a reporter—and I put myself on the path to success as well as I could.

I didn't have any evidence that my belief was right because I hadn't been a reporter yet. I hadn't followed the usual long path to distant success that everybody else believed you had to. I didn't go out blindly and pay my dues for years as commonly believed.

In general there's a giant belief that you must pay your dues and hop through all the accepted pre-ordained industry hoops before you can be a celebrity, before you can be an expert, but that's not the truth.

You can believe in yourself and transcend traditional rules and assumptions. What's most important is to believe in yourself, and regularly put yourself on the path of success. At some point the rest of the world will get the memo. Your moment arrives when your preparation meets opportunity.

In my case, just two months after I started my job at that TV station, the newscaster became suddenly sick with the flu and because I looked like I could take his place, the producer asked me if I could step in and take his place—and I did.

My positive belief launched me forward through that single crack of opportunity, and suddenly I was being who I knew I was: a real, paid reporter with a job at a TV station. I started at the top of the industry, instead of clawing my way up from the bottom, because I held the vision and took the necessary action.

Creating Room For Success

Later, when I looked back on this early success story of my career, what I found I was doing in my own mind was refusing to accept and get caught up in all the dark, negative, hopeless beliefs surrounding my career vision. Instead, I kept returning my focus to that clear, bright, open, receptive state of mind that I showed you how to enter into earlier in this chapter. That inner procedure that I described represents my long-term distillation of the natural success process I spontaneously discovered.

I might mention that many people before and after me have made the same basic discovery, and then created various approaches for entering into this special quality of consciousness where old negative beliefs have no power over our emerging passion and vision.

Voice Activation
Exercise #5

A Process for Mastering Your Self-Talk

Because this process is so important to helping you with your passion to advance in your career, let me guide you through the experience again to end this chapter. I strongly recommend that you begin to memorize the process, learn it by heart, and make it your own. Summon enough discipline to where you practice the process twice a day for a week, and it will be with you as a primer.

Take five minutes entirely off... sit quietly for a moment, tune into your breathing ... and then begin to imagine that you're walking into a room, perhaps in an apartment or home. Notice what's in the room. Are the windows open or closed? Are there drapes? Is it light or dark? Is there furniture in the room? People? What's going on in the room?

In your imagination, focus on each one of the objects that you notice in the room, one after the other, and just ... dis-

solve it. *Use a little bit of magic in your imagination and continue to dissolve all the different things in the room so that step by step the room becomes clearer and clearer and clearer and clearer ... until it's completely empty.*

Now go and open the drapes and blinds, *and open wide all the windows, so that more and more light comes into the empty room, and it becomes brighter and brighter.*

Pause, and in your imagination sit quietly on the floor ... and fully enjoy and embrace the clean clear empty brightness and the air you're breathing. *Let your heart and soul and mind become more and more filled with this wonderful brightness and clarity ... and be open to receive whatever insights and new feelings and ideas come flowing into you ...*

Pause and Experience

"Turning up our light in

the presence of those whose light

*is dim becomes **the difference that***

makes the difference."

Eric Allenbaugh

Identifying Your Gifts and Expert Voice

In this chapter I'd like to explore with you how to identify your strongest talents and capabilities based on your life story to date—and also help you build more confidence and authenticity as a noted expert in your field.

Your expertise and choice of profession or life focus has obviously emerged from your unique experience, from your all-important life story and accumulation of knowledge. Your story is in reality many stories. Every facet of your life contributes to the complete expression of you. By looking at each area independently, you can gain a deep understanding of your unique expertise and the path your development. Take a moment and write only these stories. They provide insight into what drives your forward and refine your vision.

Your Money Story

For instance, how has your personal story unfolded in relationship to that often-over-riding dimension called financial well-being—stated bluntly, money? Did you come from a wealthy family, or did you grow up with money issues straining your family? Does money come easily to you, or have you always struggled to make ends meet? Do you enjoy making and spending money, or do you have negative attitudes that stand in the way of financial success and well-being?

Pause a few moments, and reflect on these questions about your money story and your present money situation. How does your inherent need for ample money integrate into or battle against your desire and calling to be of service in the world with your particular talent and knowledge?

Your Relationship Story

Now consider your relationship story. Is it easy for you to meet new people and establish meaningful relationships, or are your relationships usually shallow, difficult to maintain, and short-lived? Have your relationships usually brought you joy? Have they brought you peace? Have you learned important lessons and insights through relating? And right now, how do your existing relationships mesh with your intent to become a successful expert?

Again, pause and focus for a moment on your breathing and whole-body presence. Breathe into any feelings you find in your face ... your throat ... your heart ... your belly and deep down. Quietly reflect upon the people you have known and loved and on what you've learned about human nature and what you're yearning for in relationships right now.

Your Health Story

Now, what's your health story? Were you born with good health and vitality, or has wellness tended to elude you and health problems dominate your life story? Were you born with radiant energy and physical gifts, or where you perhaps born with a disability? Have you learned important realizations about health and wellness that you now want to teach to others?

It's a very powerful practice to regularly pause and look back at your life story as seen from different perspectives and themes to see where your current passion and focus has emerged from in your life. You can also explore your past history from the point of view of dangers you have faced and overcome, teachers who have impressed you with their own experience and knowledge, situations that pushed you beyond your limits into new realization and power.

Once more take time, put this book aside, and look inward to memories of your life, this time related to your physical body and its ongoing lessons and your present readiness to step forward in your physical body in front of the cameras and audiences and clients and be a physical embodiment of what you are teaching. Are you ready?

As the weeks go by, I challenge you to discipline yourself each week to write down your life story from a different perspective, really dig into the sources of your present knowledge and expertise.

You have accumulated knowledge and experience and insight of high value, because you have lived through experiences that provoked growth and struggle and breakthrough. And the more deeply you're grounded in your own story, without judgment or hyperbole, the stronger your presence is going to be as you now go forth in your life work.

Your Reality Resume

What are the credentials that make you an expert on a particular topic? Is it your education, certifications, and training? What about your life experience? Your life experience also plays a crucial role in your ability to lead others. The combination of your life experience and your education is the basis for your **Reality Resume**. Many of you are held up in stepping forward because you believe you do not have the correct education or credentials, when in fact your life experience is deep and rich and full of transformational insights not learned in a classroom. Take a moment, create your reality resume, and own the fact that you have the skills, knowledge, and experience to be the recognized leader in your field. You are already an expert in your life's purpose!

Fully Recognize Your Gifts

In the process of moving through your life story from various angles and themes, you'll begin to find one or two particular angles that most strongly evoke a realization of your natural talents, acquired knowledge, and expertise that are ready for public expression. I call this your **FINGERPRINT EXPERTISE**. In this space of unique expression there is virtually no competition. You'll realize that you're right now in the process of extracting your unique wisdom into a format that you can present to the world as a genuine expert.

Let's say, for example, that you had been born with a disability that you needed to overcome. It might have been something like a learning disability, ADHD, or perhaps it was a physical disability. Maybe you had trouble learning to walk when you were younger, and you had to face that challenge and move through it—learning life lessons all the while. You developed a deep sense of determination and persistence and were able to transcend this disability that you were born with.

And now you're able to help others muster the same inner strengths to overcome their particular problems. You know the path from personal experience, and so you can teach it to others with power and clarity and compassion.

The archives of awareness. Most people don't often look back and take a journey of awareness into the archives of their lives to see what gifts they accumulated and cultivated through their lifetime of experience. But when we do discipline our minds to pause and reflect, to remember and identify, and acknowledge our gifts and talents, then we can begin to develop our gifts and knowledge into formats that we can teach to the world.

Build a Solid Foundation

We all have one, two, three, four, five things that we really shine with, that the world needs, and that we can employ to sustain us professionally and fulfill us at deeper levels as we share our gifts. My challenge to you is to honestly identify those natural or hard-learned gifts—and establish them as a foundation on which to cultivate what your philosophy is about life. Ground yourself in your own experiential knowledge and wisdom. Trust your own life.

Only through self-reflection can you come to discover who you really are deep-down. And you need this self-knowledge in order to stand before other people with authority, and teach them what you know of life.

Both your environment and your nature are always working together to mold and inform your philosophy, your beliefs, your assumptions, and your life intent. You need to reflect and know yourself well enough to know, in essence, what your bottom line is. What is your highest aspiration? What is your driving intent and life calling?

Once you recognize the truth of something, you can no longer stay idle and unresponsive to your calling. You will have a moment inside your life, if you haven't already, where you'll experience an epiphany. You'll just say, "Yes! I know what is my truth. I see the path before me. This is what I need to do next. I can't tell when or where specific things are going to happen, but I can feel it coming, feel the flow on a cellular level, on a soul level."

You'll say, "This is my moment." And every minute from the point of that recognition, if you don't follow that realization with intent and action and begin to implement the details of your life vision, you'll know that you're not stepping up to the plate. The solution is to drop into your foundation, once again clarify your intent, and get back on the path.

Over and over, you'll find that instead of saying yes to your life, you are saying no. Every day that you say no, know that you are failing to tap your own life-force energy; you are letting life pass you by.

You'll know when you are leaking life-force energy, because symptoms will begin to appear. At first you'll notice that you're having more and more bad days at work, or maybe you're caught up in a chronic cold that just won't let go—or some other negativity shows up in your life that dominates your attention.

And as time goes on when you're still saying no to life and caught up in fears of failure and inadequacy, your relationships will begin to deteriorate, or you'll start to cultivate health problems. As your energy hemorrhage continues, you will manifest some level of emotional, mental, or physical illness.

And as we talked about earlier, you'll find in this situation that you're being consumed in negative self-talk, where your ego is contracted, afraid, and judgmental and saying, "I just can't do this, I can't step forth and assume my true leading role."

The main problem with our ego's attitudes is that they're based on the fear of failure, rather on the thrill of trying. And fear of failure freezes any positive advancement.

Of course you're going to make mistakes as you venture out into the world with your particular talent and knowledge. But if you have negative one-liners in your mind that think the world will end if you goof and flop and are seen as less than perfect, then you'll never be able to advance on your path, because you are just sitting fearfully on the sidelines, not playing the game at all.

Redefining the Expert

The ego all too often has an entirely erroneous notion of what an expert is really like. The ego assumes that you must pretend to know everything there is to know about your topic before you can step forth and proclaim your readiness to be of help in the world. Because the ego is afraid, it tries to pretend to be perfect.

The true expert. In reality, the true expert is the person who has already learned a lot about a topic from experience but also continually returns to the role of beginner.

This is the only way to continue to become more and more masterful in your field—you must regularly become humble and open to receive new experience and knowledge. You're already the expert right now. What's required is to make this quantum jump to where one minute you perceive yourself as being this incredible expert in life, and then you move to the next level where you find yourself a beginner in completely new waters. This is how you continue to advance, and your ego needs to learn this lesson and honor it rather than fearing it.

Even if you are sixty years old, what's being asked of you by the higher definition of expert is that you regularly become like a free young child, fresh and observant in your mind and attitude, open to new experience that will bring you yet deeper into the infinite realms of life.

In essence, you must be confident enough in your current level of expertise to be willing to be seen as a beginner in the next round of learning. Then people will truly honor and respect and trust you to lead them because you're constantly evolving and going deeper into your life theme.

You are always more than you realize you are. If you trust Spirit in your life, you are continually receiving new insights. You are attaining a new perspective. You are waking up to a brighter vision of life. This is how you allow more wholeness to come in.

Chances are very high that you have not yet expressed your whole self authentically in the big arena. You have expressed the part of yourself that you thought was acceptable and valid and safe; that's the ego trying to be cautious and safe and accepted.

Each time you're in front of the camera, in front of an audience, or talking one on one with someone, if you're humble and open and curious and receptive, something new can happen, a deeper truth can emerge through you. You can be a beginner, and still be a leader.

Leap Of Faith

I remember going through a big change of my life. I had written a self-published book, *Mystic in a Minivan*, and had a small practice as an intuitive life coach. I was trying to build a wellness center. I had three small kids and had just gone through a divorce, and in the process, I'd lost pretty much everything. With my house in foreclosure and my car repossessed, I'd been thrown literally back to scratch. I had worked in TV news for ten years, but that was now behind me. So here I was, in a radical mid-life reinvention.

I admit that as an aspiring expert back then I was uncertain of myself and needing to somehow let go of the past and step out in a big new way. My whole foundation that I thought was going to sustain me had been suddenly swept away. I was in a complete life meltdown.

I had a general new game plan—I was going to totally push aside and reject my old life, run my wellness center, and promote my spiritual book. But what I found was that my true success, my true path, my true expression of expertise required a combination of the new person that I was becoming and the expertise that I already had.

This integration of the pre-existing you, and the brand-new you isn't easy, but that's where you'll find your opportunity. A lot of us make our lenses so small, so tight, that we miss the bigger aspect of ourselves because we think that our past experience is no longer valid.

As I mentioned before, it's not possible to leap into the unknown future on just one leg and emerge as a new you. It is essential to honor value and take full advantage of all our past experiences because that's really who we are in terms of knowledge and expertise.

Yet at some point we do need to let go of the old identity we've built up and emerge into a new creation, take the leap entirely beyond our ever-cautious ego identity, and set ourselves free to be spontaneously who we are becoming, in each new moment with each new breath.

Drop-kicked by the Universe

I have found in my work as an intuitive life coach, with all the different people that I've worked with, that in order to bring forth one's highest offering to the world, at some point a leap of faith is absolutely required. You must let go of the known of the past and have faith in the rebirthing process that's propelling you into your new role. It's like the universe saying, "Do you trust me?"

With full trust in the universe and its perfect unfolding, we can make the leap and bridge the life that we're living and the bigger one we're meant to live.

When we're meant to take a leap, when we're ready to move forward, first we get a tap—let's go, it's time to move. Then we start to get a little bit of the nudge. And then if we don't respond to the universe nudging us, it starts to get louder. Events around us say, "Excuse me, hello? Can I have your attention? Let's get moving!" And then it gets even louder, where destiny is shouting at us, "Get on with your life!"

If we still don't go into action, if our ego is playing dumb and blind out of fear, then conflicts start to come into our life all around us, everything starts to fall apart in our lives, and if we're still afraid to act we say, "Oh it's a terrible storm, it is the worst possible time ever to live." But actually, all the storm is doing is creating provocations to get you moving.

Eventually, if you get to a point where you still refuse to leave the old and leap into the new, you get what I call drop-kicked by the universe, where you are literally kicked out of the space that you're hanging onto because it's no longer habitable by your spirit.

Change is constant. If we resist, we build up pressure until we can't hold on any longer, and then we experience a personal earthquake. Obviously it's wiser to respond to change and transformation without so much resistance.

From Chaos to Homeostasis

One of my favorite writers, Margaret Wheatley, speaks often about the chaos in life and how that chaos can naturally shift our lives into renewed balance and homeostasis. We see this happen regularly in biology, and it happens regularly in peoples" life stories as well—where there's a resistance to change, that leads all the way to the state of complete chaos because everything gets uprooted.

A person or situation becomes entrenched, needs to hold on to the known, doesn't want to let go, doesn't realize that resistance to change is futile.

My epiphany. When my life was completely falling apart, I had a sudden epiphany. I knew at least theoretically that our thoughts attract our reality. My epiphany was that I should stop fixating on the outside chaos in my life—finances obliterated, relationships crashing—and start paying attention to my own thoughts. So I started a daily practice of setting aside time and space to look at the chaotic thoughts in my own mind.

What happened to me is what happens to everyone when they begin to simply watch their own thoughts in action. At first all I saw was the chaos, the anger thrashing wildly about, the anxiety trembling and rigid, the confusion numbing my decisions, the depression sapping my life force.

But through pausing regularly and breathing into my own internal chaos, and beginning to accept that chaos, something started changing for the better. One of the first things that happened was that I found I could experience a bit of distance from the chaos in my mind. I gained a little perspective, and stopped identifying so completely with the wild thrashing about of my ego mind.

And then, step by step, I found thoughts appearing deep down in the chaos that weren't chaotic at all. A sudden, clear, wise insight would pop into my mind and then disappear again. Then a similar thought would rise up once more, not even quite a thought, just a subtle feeling, a sense of possibility, of newness emerging in the midst of the chaos.

I found myself going to a bookstore and being drawn to a particular text, buying it and taking it home and digesting it, and feeding myself not only with the spiritual wisdom of the ages, but also with new insights from research in psychology. In general, I began to feel like I was being guided by a force or intelligence that I couldn't understand, but that I could increasingly sense and trust.

During the next months I opened up my intuitive channels, lost 40 pounds ... and wrote a book. Within a year after I started to listen to the chaos in my mind, gain merciful distance from that chaos, and tuned into my deeper presence and guidance, my life was transformed from the inside out. And yes, the outside chaos also retreated as my life found new balance and expression.

What happened as a result of moving through that wild chaotic period was that I shifted into a different expression, not a higher nor a lower, just a more full expression of who I am. I became more observant and acceptant of who I was and of what I had done in the past, and brought back into my self all those pieces of myself I'd rejected.

In doing this, and opening up to trust all the newness, I became bigger, radiant, and more full. And just like a snake shedding its skin when it outgrows it, I appeared new to the world.

If we become conscious of the process we're going through when chaos seems to take over our lives, I've found that we can choose how we perceive that situation—we can react and see the turmoil as really frightening and painful and say, "Oh no, this is terrible, I'm losing something here," or we can respond and welcome the change, and actually feel joyful and thankful in the midst of the illusion of chaos.

Your Authentic Voice

One of the great values in going through tumultuous times and staying conscious throughout is that you begin to discover your deeper core of being, and in this process you tap into your authentic voice. People will hear in your voice that you've really lived life and know what you're talking about first-hand, and they will trust you to guide them.

When you speak with an authentic voice, no matter what you speak about or how you connect, every time you open your mouth your words come from your heart and your soul.

Living an authentic life means not resisting what's happening, nor defying your destiny, nor denying the truth. This might seem to be a difficult path to follow, but in fact when you flow with the river, rather than fighting against it, life becomes almost effortless.

Each new moment is God's creation emerging in time and space. To be authentic means to be a part of the greater reality of this moment. And when you live in this authentic place, expressing the truth you find within you and around you each moment, then your voice is naturally authentic.

If you look at your evolution as a loss, if you mourn for the past and cling to the known, you lose your connection with Spirit, which is manifesting in the present moment, and your voice will lose its authentic ring. Your vision will become blurred, and you'll see chaos everywhere around you—even though from a different perspective, the world around you is perfect, balanced, and in harmony.

Change is often seen as chaos—but resisting change is what generates chaos. As you evolve, you become more of who you're meant to be.

I say go ahead and trust change, live the life you are destined to live. I believe that every one of us has a divine assignment; we've entered our life with a pre-ordained path of expression. Our journey is to remember this assignment and purpose (once discovered) and be unwavering in its pursuit.

Voice Activation

Exercise #6

Four Affirmations

Here's a set of intent statements that will aim your life in the direction you want to move in. If you return and move through this four-step process often, you'll accelerate your evolution.

Get comfortable and turn your attention away from your busy day. Feel the air flowing in and out of your nose and the movements in your chest and belly as you breathe ... say to yourself:

"I honor, accept, and integrate all of my past."
"I embrace the change happening in my life."
"I choose to express my real authentic self."
"I feel confident that my life is unfolding perfectly."

Pause and Experience

"You are already an expert in

*your life's **purpose**."*

Kristen White

"Your time is limited, so don't waste it

living someone else's life. Don't be trapped

by dogma, which is living with the

results of other people's thinking. Don't let the

noise of other's opinions drown out

your own inner voice. And most important,

have the courage to follow your heart

and intuition. They somehow

already know what you truly want to become.

Everything else is secondary.*"*

Steve Jobs

Crafting Your Message to the World

Many of us have considerable trouble integrating our passion to teach our message with our need to support ourselves and our loved ones financially. In this chapter I'd like to share with you my understanding of how money and message, when both are approached in the proper spirit, not only work together but are often essential to reinforce each other.

What are the elements of your story that translate into a viable business?

You can work on this in depth with my FREE BONUS found on page 203. Find the Money in Your Message.

Connect with their pain. From my personal and professional understanding, the most powerful way to find wealth in your life-calling is to learn how to speak a message that connects directly with the primary points of pain of your ideal reader, audience, consumer, or client.

What I mean by this is that our planet is mostly populated with seekers—people who have key issues, fears, challenges, desires, or obstacles that they want to somehow transcend. These people are willing to invest their resources, their time, and their energy to follow an expert on a path that provides relief, solutions, and abatement of their core challenges.

Very often, people are resisting change in their lives. As we said before, they are fighting against the reality of their evolution, and as a result, they are generating pain in their lives, of one form or another, with countless symptoms.

You need to understand your ideal customer at the level of the pain they're currently feeling, and help them realistically move forward and relieve that pain.

Often, people are looking hungrily for inspiration to solve problems that they don't even realize they have. They're on a journey of deep inner growth, and awakening and realization, but all they see are the bundles of pain along the path, not the path itself.

You have walked that same path, and so you can have compassion for people you find struggling on the same path, and you can offer realistic help and guidance because you have been there yourself. You know how to advance on the path with finesse and joy, rather than with sprained psychic ankles and banged shins.

People staggering along the path with their eyes mostly closed are of course going to experience pain. You can teach them how to open their eyes and walk consciously through their emerging moment. And once you gain their trust, you can guide them not only away from painful pitfalls ahead of them in life, but also toward bright opportunities for transcendence.

You are not there just to help them solve their identified problems; you are there to help them find creative solutions that stretch them beyond who they presently think they are.

And when you are of significant help and gain a client's trust as their guide, they will respond with deep, lasting loyalty, financially and otherwise, as they learn to invest in their own lives, by offering you financial support as you guide them. Everyone wins.

Your challenge is to understand a general life situation very well, through your own experience, and then write a book, create coaching programs, craft products, and so forth, that will guide and allow people to expand beyond a restriction or a point of pain that they are in right now. And in so doing, quite naturally your prosperity will flow in from the shift you facilitate.

Gaining Message Clarity

There are a vast number of people out there who are seriously hurting, who are avidly seeking, and they will definitely listen to your message, but only if it's composed and presented in the proper format.

Let's take a sample message in and break it into a couple of different pieces.

Step One: Deliver a one-liner that's an attention grabber: When I worked in news I followed the Nielsen ratings and polling statistics. They're one of the main companies that studies audience behaviors, and they found in their research that most people have an attention span of only about 15 seconds. What that equates to is one sentence. In business that sentence is called "the elevator speech," Hollywood calls it "the high concept," and in the media biz it's known as "the sound bite."

That one sentence must clarify your credential in your expertise. It must also at least hint at your benefit-oriented solution, contribution, or expertise. And it needs to be so simple, focused, and clear that once someone hears it, they can easily repeat it back to you. That is your first sentence; that is your attention grabber.

Step Two: State your theme succinctly in the format of a question: Questions always evoke attention, and this question needs to continue to hold attention following your one-liner attention-grabber. For example, I recently worked with a client who wrote a book on solving high cholesterol naturally. So her question would be, "Have you or someone you love been diagnosed with high cholesterol?"

Step Three : Establish the urgent need for effective action: Related to our example, you need to point out that if you don't treat the condition, you could have serious health ramifications and die—but if you treat the condition with regular drugs, you could have serious health ramifications in reaction to the drugs. Is there a better solution to the problem?

Step Four: State your claim to authority in your chosen area: The next piece relates to how you claim your expertise. This is a sentence where you introduce yourself. For instance,

"Hi, my name is _____, I'm an expert in this area and this is my experience (list your experience) _____ and these are my results (tell a few quick customer results) _____."

"I help _____ (your ideal client), resolve their problem with _____ (a specific issue)."

You can then add more details as in the following examples.

"I help _____ (your ideal client), find peace and a better life (describe) _____, by following these steps (list the steps) _____, to relieve these problems in this (time frame.)"

Make this statement of your credentials as short as possible, don't dwell on yourself.

Step Five : Explain very briefly how you can help to solve their problem: Always return the focus of attention to the client and their problem as soon as you can, rather than staying focused on yourself. They want solutions, not resumes. They need to know if and how you can solve their problem.

As you are talking to potential clients, realize they are busy searching through the information for clues you have a solution to their problem.

They are thinking to themselves, "Do I connect with you? Are you the one person who has what I need to solve my pain, my frustration, and end my struggle?"

Step Six: State your call to action: The conclusion of your message is probably the most important step because it's the implied command; it's the call to action. A lot of people are great at grabbing a potential client's attention, they are wonderful at claiming their expertise and outlining their solution to a problem, but then they don't call for action.

If you're doing your message delivery online, for instance, this is where you say, "Go to my opt-in box on the top of my website and enter your name and e-mail address so that I can give you this transformational content."

These are the six steps in a successful message. Of course there are many variations on the theme, but in general you will want to work with these guidelines and create a flow that sounds effortless, friendly, and authoritative.

Envisioning Your Optimum Client

If you're new to writing this type of short sales document or presentation, here's a good way to start the process of actually finding your message and then attracting the money associated with that message.

Take the time and effort required to make an extensive list of all the attributes of your ideal client. It's best to pick one person; choose your favorite client. The person can be paid or unpaid; it doesn't matter. But start with one person because it is much more effective in your marketing to speak to one person than to try to speak to a hundred people or a thousand people at the same time.

Just jump in and start to define who they are:

   1. What do they do for a living?

2. What are they passionate about?

3. Are they married? What's their relationship status?

4. What's their age? Do they act young or old?

5. Are they male or female?

6. Are they a high achiever?

7. What are their primary wants, needs, and desires?

8. What are their primary frustrations?

9. Are they optimistic or pessimistic?

10. Are they happy or sad?

11. Are they healthy or sickly?

12. Are they wealthy or poor?

Now continue listing all the aspects of this person that come to mind until you have a full clear description of that person.

Focus on the client. In crafting your message, and talking to pro-spective clients in general, 90% of your language should start with the word "you" and an outward focus message to your client. The big mistake, for instance, that many people make on their web-sites is they're always talking about me: "I did this. I'm an expert at that. I've written many books. I've done these things." Nobody cares about that. They want to know how you impact them.

In all of your marketing, your copy on your website, in your book, and the presentations that you make, if you're not using the word "you" or something very specific (even their name, there's a lot of technology out there that allows you to insert someone's name), then you're probably losing your audience.

If 90% of your content—even on your own personal website—isn't about your ideal customer, then your content needs a complete overhaul.

List the benefits of working with you. Another way of fine-tuning your message is by making a list of all the ways in which your coaching, product, or service benefits someone. Again, make a real list of every particular way your offering can reduce pain of any kind, can solve a problem or relieve a particular negative situation, and can generate a particular positive outcome that people yearn for.

Take time with this list. Play with it over several days. This will help your mind begin to create a clear image of the people you're seeking through your message. You must tune into a person and prepare to talk directly to them, work with them, get to know them, and be of service to them. Then you will begin to connect and perform valuable service, and the natural response from your clients will be both personal and financial.

What does an enlightened expert need to do or say in order to draw their ideal customer into the transformational offer that they're making?

Asking The Right Questions

People have confusion and problems that they are seeking to unravel. If they could ask the right questions, they'd know how to unravel their problems step by step. An expert is someone who knows the right questions to ask that lead to awareness.

Your job as the expert coach is to ask powerful questions. One powerful question leads to the next question. Usually these are not questions that can be answered with a yes or no; these are questions that are more expansive and require a greater contribution from the person being queried.

When you're a enlightened expert, you are operating from this wise coach capacity. You are constantly asking a series of powerful questions of your audience, questions that invite them to respond, interact, and exchange with you.

Creating and asking powerful questions is truly an art form in and of itself. Questions can show up in a personal exchange with your client, they can drop inside your marketing, and they can be used in your videos. Powerful questioning is open ended. A simple test to see if your questions fall within this category is that they can't be answered with a yes or a no.

Usually questions start out more general. What happened this week that made you happy? What are you celebrating this week? Then they can go deeper. For instance, asking about the vision that you hold for yourself six months or a year from now. Who would you be if a particular belief system that you adhere to were to change?

There are hundreds of powerful questions. Often the best questions emerge intuitively in the moment. This goes back to what we were talking about earlier in this book, of maintaining a bright empty space inside yourself, a space that you can hold open for the people that you're working with.

In this empty clear space, with the windows of wisdom and insight wide open and your own personal stuff mostly gone, you can come from a place of deep listening. And while you listen, you can begin to intuit what's coming up for the person, and then a natural intuitive flow will come to your mind about what to ask next, and you express it.

The only thing that you need to remember as you open to intuitive guidance during a session of this sort is to make sure that the questions are open ended and that they lead deeper and deeper into eliciting sudden epiphanies and realizations of truth in the person who's answering them.

Empathic Listening

Questions are designed to elicit information. How you listen to that response to your question is as important as being able to ask a powerful question. Most people are hungering for someone to listen without judging or interfering with them. The foundation of traditional therapy, for which people have been paying great amounts for many decades now, is the therapist's ability to enter into a state of empathic listening.

If you want to be a successful guide, it is essential to actively develop your capacity to listen well so that clients feel truly heard.

A lot of us assume that we're good listeners, and we are shocked to realize that we're not. What too often happens is that in conversations we hear somebody else talking, but we're already formulating what we're going to say back to them.

We're already crafting our next point, and therefore our attention is not fully on the speaker. We pretend to be in receive mode, but we're already in broadcast mode.

So one of the major steps in becoming a successful expert is learning how to be genuinely "there" for your client, after you ask a question. And to accomplish this fine art of being a patient focused listener, when you learn what I've been talking about throughout this book—the art of quieting your mind—you'll become a pure channel that accurately receives and decodes information from the souls of others.

Voice Activation
Exercise #7

Listening Empathically

Here are some secret tips for finessing this art of listening empathically. This technique is found throughout a number of books by a good friend of mine, John Selby, and he's given me permission to share this technique with you here, as found especially in his book, *Listening With Empathy*.

When you are with a client and asking key questions that elicit a deep response (or actually any time when you're listening to anyone at all), *here's a specific process that focuses your attention fully away from the chatter in your own mind into a quality of receptivity in which you truly hear what the other is saying, with no judgment, no reaction, no preparation for what you're going to say.*

After asking your question, in which you are requesting a response from the other person, immediately shift your own focus of attention to the sensation of the air flowing in and out of your nose right now ... tune out your thoughts in your brain, and tune into the actual feeling being generated by the flow of air in and out of your nose.

Now expand your awareness to include the movements in your chest and belly as you breathe ... and continue expanding so that you become aware of your whole body at once, here in this present moment.

And expand your awareness now to include the room that you're in and the person in the room with you that you're listening to ... don't even try to understand or process what your client is saying, just take in the words as you continue staying aware of your breathing experience.

Be sure to hold your breathing experience as most important and the sound of your client's voice second in importance ... and let your whole being be entirely and enjoyably filled with the present moment experience of listening and breathing, breathing and listening.

As you breathe and listen, you'll also find yourself with an inner feeling of emotion ... breathe into this feeling of emotion and let it expand and change in response to what your client is saying ... let yourself actually be touched by what your client is sharing with you ... experience genuine empathic communication as you listen.

Allow your client to talk until he or she has expressed everything that's ready to emerge from them ... let your conscious breathing sustain your ongoing patience until your client becomes quiet ... and still, keep your full attention on your breathing ...

your feelings … your awareness of your client's presence in the room with you.

In the midst of this quiet communion of shared space with your client, be open to receive a flow of insight into what you might now express … *see if a new question comes to mind from your deeper inner source of wisdom and guidance … and when you're ready, without slipping into regular analytical judgmental thought at all, see what new question emerges ... and express this question.*

And again, you are free to tune into the air flowing in and out of your nose … *the movements in your chest and belly as you breathe … your whole-body presence in the emerging moment and the flow of communication emerging from your client, in response to your question.*

This is what it's like to be "in the zone" as you do your work. *You are actually "doing" nothing. You are "being" present and allowing your client to discover their unique inner truth and path and healing … and for this service, you will always be more than amply valued and rewarded!*

Pause and Experience

"If you can't explain it to

*a six year old, **you don't understand***

it yourself."

Albert Einstein

"I've learned that people

will forget what you said, people will

forget what you did,

*but people will never forget **how you***

***made them feel**."*

Maya Angelou

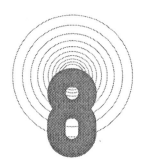

Scripting Your Voice and Persona

Every successful coach or teacher of any kind works with **a basic script that they say over and over in many different ways**. You're teaching a particular path and that includes a spoken oath, and even though you will express your script in many different variations, you're still working from that basic script.

Your script will ideally consist of around 40% questions developed and employed as we discussed in the last chapter. But there is also much more that goes into a successful script beyond powerful questioning.

What you're doing with your client is engaging them in a conversation. You are wanting to talk with your client about their needs, wishes, and problems. And to activate that dialog, you're going to first need to ask questions that elicit responses.

Content marketing, not interruption marketing. What's going on with marketing today is a focus on content marketing, on what we call reflection marketing. It's a process of tapping authentically into the emotions, the journey, and the needs and wishes of the people out there that you are meant to serve. A question is asked and a consumer reflects on the response.

This is in direct contrast to the marketing approach that used to be popular, which was interruption marketing. It went like this: "Let me interrupt you with what I have to offer. Let me tell you why I'm an expert and what I have to sell." That doesn't work anymore.

Relevance and engagement. People want relevance, and they want engagement. They want invitations to participate in what's going on for their benefit. They want their questions answered. That's why they go to a reference site like Wikipedia or WebMD to find information. But all too often they don't get their questions adequately answered.

People always have more questions, and they will embrace the expert that asks them the right questions and allows them to respond. Then and only then will they listen to what you have to say and listen to the rest of your script.

They're going to go to search for answers to their questions online, and at some point you're going to appear, and ask the right questions, then listen to their story and guide them where they want, need, and deserve to go in their lives.

I believe that one of the most important things that an Enlightened Expert can have in place is a really strong personal blog with personal videos. Inside this blog, they can post questions and respond with articles that have powerful content, and build credibility and relationship with the people who respond and engage.

As experts in this new arena, we're not putting ourselves out there as, "I know everything, so just do what I say." It's more like, "Here I am, ready to walk alongside you, listen to your story, and share what I know."

Sustaining Relationships with Clients

Let's talk a bit about the long-term. How can you develop a script that enables you to continue working with a client beyond the short-term resolution of a particular question into a more con-tinual supportive relationship?

Your online and offline presence. Perhaps the most important answer to this question, which might startle some of you, is that the web is not the whole story. There are a lot of touch points out there, a lot of places where we can have a connection with somebody, but then fleetingly it goes away. We don't ever see or hear from that person again; they don't respond to your emails or unsubscribe. What can you do to develop stronger, more meaningful relationships with your potential customers?

In my opinion, based on what I've observed: If you are relying exclusively on creating online relationships and using social media as a way to get customers and become a Enlightened Expert, you will probably fail. You must travel, speak on stages, and meet people face to face.

I feel quite strongly on this point. It's important to recognize that social media is just another tool of communication. Just like a cell phone, a land line, or any other non-local means of communication, nothing replaces getting out there and speaking in person, going physically to events, working one-on-one with clients, and connecting with people by putting yourself personally in the community.

You can of course then retreat, follow-up, and run your business online. Certainly maintain a polished and sustained online image and presence, but put the muscle in your offline presence. That's how you build and sustain meaningful client relationships.

Script Variations

There are a lot of different opportunities, online and offline, for your basic script. You can modify your script to suit each of these moments, and you can deliver it as printed content, online written content, audio content, and video content.

Short introductory script. Certainly you'll need a short introductory version of your script that welcomes people to participate on your site. This will talk a bit about your understanding of your ideal customers' pain, establish your expertise, and briefly mention your system and solution, which delivers your call to action and invites them to opt in. I recommend that most of your scripted messages be no longer than two minutes, especially for a video, because people's attention spans are highly limited.

Multiple content bits. You are better off to have several smaller bits of content than to present massive five-minute scripts. Viewers have been so conditioned to the sound bite, to fast changes in media content, that if you go on and on without a break, people just don't hear you. You may like the sound of your voice for five minutes, but for everyone else two minutes is about the max. I keep most scripts under a minute; you can cover a lot of ground in a minute, especially if you get to the point. This rule of thumb also applies to a media interview. Do your best not to talk non-stop for more than four or five sentences. Then pause and let the other person respond.

You want to quickly express compassion, show your expertise, and present your call to action. Ask them to opt in. Tell them to visit your Facebook page. Ask them to subscribe to your channel.

I view scripts as traffic lights. People encounter your business or your site, they come across a message that you've scripted, and right at that moment, when they stop to listen to you, your job is to establish that connection. You connect with them and show them that you understand where they're coming from, that you have compassion and solutions for their pain and their struggles.

Be very specific. Like a traffic light, you have them stop to listen to your intro presentation, and then give them the green light to continue deeper into your inspirational city. You're getting them used to letting you be the guide. Issue gentle commands, and help them actively engage with you by making suggestions concerning the next step of the process.

In the guide-client relationship, where you are being paid for your expertise, it is vital to clarify early on that you're comfortable in your leadership identity, and that you expect them to participate with you in that relationship. You need to step forward as the guide, the leader, the healer, the one who knows the territory and knows the process for relieving the client's confusion and pain.

Long script for on-stage presence. But be sure to vary your script and your discussion to suit the situation. When you're on stage delivering an hour-long talk, obviously you're going to have a longer version of your script. You need to study and master that format if you're going to succeed with a large audience. It's a multi-dimensional performance challenge to hold a large group's attention for that extended period of time.

Equally, when you're doing a radio interview, you must be careful not to talk too long at once, to ask questions of your audience, and to present yourself as a guide, not a know-it-all.

And when you're in that intimate one-on-one client relationship, be sure to be a good listener. Remember what we talked about earlier in this book, about making your awareness of your own breathing and inner emotional presence most important.

In whatever format you deliver your script, you must maintain inner power and presence. You must be anchored in your breathing and whole-body charisma. Then, and only then, will your voice carry forth with confidence, authenticity, and compassion, and that's what your client needs to hear.

You can work more deeply in this area by accessing my FREE BONUS found on page 203.

Does Physical Appearance Matter?

The answer to this question is definitely yes. You're not being vain or self-indulgent if you pay serious attention to how you look onstage; you're being wise and proactive. How you look sends a message, especially when you go on a TV show, but equally when you video short scripts for your website, give a lecture, or meet with a client in person. You are first and foremost a representative of your brand.

If you get an invitation to be on TV news or other media show, let go of the assumption that when you arrive in the studio there's going to be a makeup artist and hair stylist, someone to make sure you look perfect and flawless. That's a fantasy. The reality is that you put on your own makeup, do your own hair, apply your own lipstick. You put on whatever you think is going to look good, and you arrive in the lobby of the news station ready for the show.

These days the person who's interviewing you doesn't come out and greet you, that's the job of an intern. And they keep you waiting in the reception area until about two minutes before you are scheduled to be on. They then escort you directly to the studio and plant you next to the reporter or anchor who is interviewing you.

The interviewer gives you a nod and maybe waves hello—and you're LIVE. You're the focus of media attention for a two-minute segment, with the host asking you a few key questions. You'll be able to get out three sentences on each question, hopefully related to your basic script—and then they're thanking you and you're gone.

Preparation is crucial when appearing in the media. Your message, image, and offer should be clearly poised and waiting for the opportunity to be expressed to the masses.

Key On Camera Media Tips

Powder: Whether you're a man or a woman, you absolutely must wear powder for every media appearance. You need to powder your face to reduce shine. Go to any department store and buy studio powder, and make sure to apply an even layer before you go on.

If you're a man you're going to feel a little strange about this at first. But rest assured, this isn't baby powder, it's not white powder, it's makeup powder. Have them match your color tone when you buy it, and always keep a little media kit with powder nearby.

Put the powder on right before you go before the lights and camera. Why? Because chances are when you get to the studio you're going to be a little tense, and you might perspire. The hot lights will exacerbate this. And without the powder, you'll look really shiny—a shiny, sweaty expert. And this is just not a good thing. So, just say yes to powder.

Rethink Your Makeup: For women, even though you may think that the makeup that you wear is quite good for you in general, things like super harsh bright red lips, strong eyelashes, heavy eyeliner, or dark eye shadows don't translate very well on TV because of the lighting and the high-definition clarity of the TV cameras.

For TV, you want just the opposite of traditional heavy stage makeup. On TV you really want to look more natural, you want to tone down how you appear.

If possible, I recommend getting a formal studio makeover, and learn how it's done, and then you can prepare your makeup on your own for later engagements. But again, carry in that same little media makeup kit. When you get to the station a bit early, they will let you go back into the bathroom so you can freshen up your look.

Hair Preparation: As with makeup, you are usually going to be responsible for your hair when you go on TV. If you have crazy flyaway hair, it's important to know how to smooth it. Super curly hair, and designs like bows, berets, bobby pins—none of that looks good, so try to get your hair as smooth as you possibly can. And make sure you have inside that bag a small amount of hair spray so you can calm down the flyaway. You won't know what the background is going to be in the newsroom, so the smoother your hair, the better.

One of the things I learned when I was working in media is that hair length just above the shoulder tends to look a lot better on the air than hair that's really long. If you know you're going to be doing a lot of media appearances, you might want to cut your hair to shoulder length. This of course isn't essential, just something to consider.

Clothing: We all have our particular styles of clothing, and to a certain extent it's to your advantage to look distinctive, to stand out from the pack. But you'll also want to be conservative enough not to put off potential clients with your clothing.

We're living in a world where men feel compelled to dress up in their business suits and look identical in a coat and tie, especially in front of the media. You're encouraged to just take off the tie when going for an interview. Relax.

Usually, men end up putting on their suits for the media out of fear of what would happen if they violated this remarkably rigid dress code. As an expert, you must decide if you want to break out of the strictures of current male dress code for the media, or not.

The same rules apply to women. Be yourself, but don't be extreme just to make a statement. Make your statements verbally.

On a more pragmatic note, the best thing to wear on camera is plain, solid colors. The camera uses a variety of technologies for focusing, so if you have very fine woven prints, herringbone patterns, or anything of that nature, the lens will tend to dance on top of your wardrobe and it will look like there are ants crawling on your clothes.

I recommend colors in the more subtle range. I would not wear electric green, super hot pink, or really strong colors like orange. Dark colors absorb a lot of light on a TV set and the newsrooms are really bright.

You'll notice, if you pay attention, that a lot of TV anchors wear colors: reds, blues and purples, pinks, grays, greens. They're not wearing white, black, or anything that's heavily patterned or has an extreme on the color spectrum. I'm not saying that you can't wear black, but stay away from white.

Remember, in the studio they're going to need a way to attach the microphone. Hold that in mind when you choose your shirt or dress. Is it practical for attaching the mike clip and for dealing with the cord? This is an important consideration.

Make sure your clothes fit properly. If you wear glasses, have a current design. New glasses are an important part of your personal style wardrobe every year.

Also, if you're going to wear jewelry, select jewelry that is not noisy. A lot of people don't realize that they've got bracelets on and they're clinking. Try to wear jewelry that is quieter in terms of both its volume, and its presentation.

For more free tips, go to page 203 for instructions to download my FREE GIFT: 22 On-Camera Tips from the Media Pros

---

Your Inner Dress Code

Your basic script is verbal, visual, and radiant. What you say, how you look, and how you feel are all equally important. To end this chapter, let's focus on the "feeling" part—which relates how you manage your emotions onstage.

We've spoken in general about this in earlier chapters on emotional calm and presence. Here are some further notes, for when you're nervous and stepping on stage.

What Works For You: If you already have a method for getting calm and centered before stepping on a stage, then just continue to use what works for you. Many people develop early in adult life their particular strategy for calming their nerves. Sometimes the strategy is idiosyncratic but works well for them.

In The Zone: As we explored earlier, being in the zone is the best mode for presenting yourself on stage, and this quality of consciousness is best activated by:

1. **Remember to pause and take charge of your thoughts and emotions.** That's the first big step. Don't get so caught up in stress and anxiety before going on stage that you become lost in that negative panic emotion.

2. **Actively shift your focus of attention to the air flowing in and out of your nose …** expand to include your breathing down in your torso … expand again to include your whole-body presence. Breathe slowly, exhale completely, and don't get caught in tense shallow breathing that reduces your oxygen and makes you dizzy. Take charge of your breathing.

3. **Develop a self-talk routine in which you say sentences to yourself that calm you down and help you get centered in your expert identity.** Say something like, "I am the recognized expert in this field. I am calm, cool, and collected. I deliver my message easily and enjoyably."

Self-talk in this regard is huge. Usually before going on stage we find ourselves filling our minds with negative thoughts such as, "my mind is a blank, I'm going to make a fool of myself, I wish I hadn't agreed to do this, my hair's a mess, I'm so nervous I'm going to pass out," and so forth.

Instead, memorize beforehand and practice bringing to mind a set of positive thoughts that will boost you up. "I'm good in front of cameras, I love getting up and talking about what I have to offer people, the world needs what I have to give, I'm just going to be myself and shine!"

You need to learn how to release your attachment to what other people are going to think of you. Don't tell yourself that your whole future depends on your performance. Be open to allowing your life to unfold naturally.

In this spirit, see if you can let go of trying to manipulate the outcome. Say to yourself, "I am going to actually listen to the questions that they ask me and respond from my heart."

On Stage Suggestions

Many people practice such pre-stage techniques to reduce stress and stay in the zone, but when they get in front of the cameras they lose it and freak out. Instead of staying aware of their breathing and maintaining an expansive grounded whole-body presence, they panic and go mostly unconscious.

This is a basic reaction pattern of human beings. If we find ourselves in a threatening frightening situation of any kind, we either turn and run away as fast as we can, or get angry and attack or play dead and go unconscious to avoid the threat.

The reality is, you are indeed on stage. You can't run away and you can't attack, so you stop breathing, get dizzy and foggy in the head, and basically "go away" inside. This is exactly the opposite of being in the zone, and it's a media fumble when it happens during an interview.

Breathe. When you're on stage, your first challenge is to stay aware of your own breathing. This will keep you grounded in your body. It will keep your oxygen levels correct for optimum functioning. It will keep your mind quiet of negative thoughts, and it will let you enjoy the present moment.

If you stay aware of your breathing and whole body presence instead of fixating on pushing your agenda, you'll be able to smile and make heart contact with the interviewer, and then answer the question that's asked of you.

Reflect before you respond. Don't jump the gun and start answering the question right after it's asked. Instead, take in the question, reflect a moment, and directly respond to the question that they asked you. Wait two beats … one, two … stay aware of your breathing, and then answer the question. In this way you'll maintain your charisma, your whole-body presence, and your answer will radiate with power and impactful authenticity.

Be brief. Don't go on and on and on. Some people answer a question clearly, but then keep talking like a runaway freight train. Remember that most sound bites are 15 seconds.

Think sound bites. People who talk in sound bites are the people that TV shows love to invite back, so do your best to respond with no more than one or two sentences. That way, your message will be short, clear, and something that people can remember.

State your core message. Make a short, bold claim. Name a couple of primary benefits. Repeat some key point several times in the interview, in slightly different words. Drive your primary point home in a concise compression. Whatever your core message, and whatever your new system, it's your unique strategy—state it.

Tell them clearly what the positive outcome or impact will be for the people that you work with. State what you can do to relieve the pain, tell them calmly, and enjoy every minute. Then you'll shine.

Voice Activation
Exercise #8

Media-Ready Sound Bites

Take time right now to write out your short-form message. Struggle a bit if you need to until you are satisfied with the wording.

Next, practice saying your core message out loud until you've mastered your sound bite.

1. *Write out your script.*

2. *Read it aloud several times.*

3. *Read it aloud and record it and listen to it.*

4. *Have a friend ask you your media-ready questions and respond with your sound bites. Don't use notes.*

5. *Repeat the above steps and practice your sound bites every day. Especially, whenever anyone asks you the question, "So what do you do for a living?"*

Pause and Experience

"Understand and be confident

that each of us can **make a difference**

by caring and acting in small

as well as big ways."

Marian Wright Edelman

Part 3:
Media

"I don't want to end

simply having visited *this world."*

Mary Oliver

Using the Media to Amplify Your Voice

Learning how to quiet negative thoughts and believe in your true calling is an ongoing process. Please be sure to return to the first four chapters often, especially to their guided sessions that each day will take you deeper into your own inner vision and expression. A bit of discipline and practice will take you far.

Meanwhile, you will want to begin learning how to move out into the media world and practice spreading your expertise. This means learning how to interact optimally with the media.

Getting Your Story Out

In essence, you need to learn how to get your story out. This means expressing your unique talent and offering so that the outside world hears you. And this is best accomplished by learning how to connect with the media, so that the media amplifies your message and propels you into public light.

Using traditional media. There are all kinds of new media to explore. But still, one of the best ways of getting your story out is through the traditional media, and this presents a unique challenge due to the nature of how the media operates.

I spent more than a decade as a news reporter and anchor in Phoenix, West Palm Beach, and St. Louis, and I discovered early on what people don't usually realize: that the media is a separate culture entirely unto its own.

Here's the perspective from a media insider. We work every holiday, have no established off-hours, and work unusual schedules. I would routinely go to work early in the afternoon and finish at 11:30 p.m., after the end of the evening broadcast. As a single person, that was when I had open time to go out on a date, but who wants to go out on a date starting at 11:30 at night, following the newscast? Media personalities are not in the flow of society; our cycles are off. Journalists live inside the media, not in the everyday life that the rest of the world functions within.

So, understanding that fact, and also holding in mind the incredible intensity of deadlines that the media is under, it's clear that the strategies for capturing the media's attention need to be extremely intentional, crystal clear, focused, and persistent.

- ***You need a sharp, short message***

- ***You must be highly determined to get on air***

- **You must present your message as visually as possible for TV and the web**

Here is the criteria used by assignment desks in newsrooms across the country to determine if they will cover the story.

Is the story VISUAL? Can people see what's happening. Is action taking place that can be captured by a video camera?

Is the story TIMELY? Is it happening right now. Evergreen stories are rarely covered. There needs to be a compelling hook to attract the news van to your door.

Is the story EMOTIONAL? Does it have an unfolding human drama. Is there a story of tragedy, triumph, or even humor. Conflict between people or causes is almost always covered if it impacts the collective.

Is the story LOCAL? A news station in Florida will not cover a story in Oregon. The news covered by local affiliates is unfolding in the neighborhoods inside their market. The exception is a national news story.

Note also that almost everything that's handled in newsrooms is what has just happened, or what is happening instantaneously in the present news moment. But as an expert, author, speaker, or a coach, you're going to be that 5% that's not breaking news. You're already in the margin of any newsroom's attention span.

Ideally, you must breach the gap and become part of the breaking story, otherwise you're only going to get 5% instead of 95% of the attention span inside newsrooms.

Advancing Your Story for the Media

Usually, experts have been cultivating their expertise over their entire life. Experts are considered "evergreen content", so their story seems the exact opposite of timely. Often it isn't really local, nor does it have a dominant emotional charge. It may feel urgent, visual, and important to you, but in the larger picture, it's not.

Find a way to be part of the wave. What you need to do is find a way to jump on the current news wave, and adapt your message so that you fit into the ongoing media conversation.

In this light, let me share an example of someone that I worked with. Her name was Victoria, an elder care lawyer. She went through an incredible journey centered around trying to bring a baby into her same-sex relationship. She eventually achieved success in her goal by having twins through a surrogate mother.

Realizing that her deep learning experience might be of high value to other people, she took the time to research, reflect upon, write, and find a publisher for a book called *Getting To Baby.*

What next? Because she was passionate about her theme and story, she naturally wanted to attract the media, but she realized that, first of all, she should devote time to building her personal brand and generating a polished multi-media professional image.

She did quite a lot of preparation, practicing being in front of the camera and expressing her message clearly. She also learned how to radiate a natural inner brightness that looks attractive on camera. She then created really great-looking videos for her website that show her as a polished media-ready person. Additionally, she practiced her soundbites, and regularly published articles about various aspects of her journey in her blog.

In essence, she became her own media company, using the latest technology to publish, produce, and promote her unique content, and in so doing, she became ready for media coverage.

She started to become more educated about how to become visible to TV shows. She developed an excellent press kit, and made a list of which shows would be best to appear on.

One day, she happened to notice that *The View*, an ideal national

TV show for presenting her book, was going to be covering the theme of surrogacy because Elton John had recently brought a baby into his life through a surrogate mother. Two days before the show was to focus on that theme, Victoria sent out her own press release—she saw a wave she wanted to catch and paddled out to be in position.

Literally overnight she got 2,900 media pickups of her article. And so it came to pass, seemingly by pure coincidence but actually through a lot of strategic thinking and action, that she found herself exactly congruent with the same thematic wave that Elton John and *The View* were generating. And thus a mostly unknown lawyer got her book, *Getting To Baby* right smack in the middle, riding an emotionally charged present-moment media wave to high visibility and lasting success.

How To Generate Opportunity

Sometimes it might seem almost like looking for a needle in a haystack as you search for a media wave to ride with your theme. You will need to tune into that perceptive frequency that suddenly says, "Oh, I see an opportunity here; I see a new place where I can participate and succeed."

Know your market and your message. When you want to actively develop and promote your brand, it's essential to be very clear on two things:

- *Know who your market is.*
- *Know what your message is.*

Who is your market? In the beginning, it's going to be your ideal client. And you must get to know that ideal client inside and out: what is their pain, what is their interest, how are they trying to solve their pressing problem, and do you have a solution to their problem?

What is your message to them? When you know what your ideal client is thinking and feeling, then you will quite clearly know what your message is. And you naturally become clear on what your key focus topics are.

Victoria, for example, knew that her focus theme was anything that related to surrogacy, everything that had to do with the yearning to have a child, and with the ins and outs of fertility treatment, surrogacy, and with having a baby adoption.

It needs to be the same basic scenario for you. Get your topic clear and your client clear, and you are ready. You can put a Google alert on your e-mail so you get notified anytime someone talks about your topic in the news. You must make it part of your daily routine to do a quick scan of what's going on in the world.

These are key requirements for being prepared to leap rapidly when a new opportunity arises for action, for identifying media waves to ride, and for putting yourself on the path for success. There are a lot of great services out there to chose from, and any of them can absolutely work for you. What I would encourage you to do is plan on catching and riding more than one media wave. One blast to the media is not enough. And whatever service you pick, stay with it and publish through it consistently. This goes back to what I was talking about earlier, becoming your own media publishing company.

Become the media. The strongest and best way to attract the media is to become the media, and that's an opportunity that has never existed in our world—until recently.

Everyone who wants to can now become a media outlet, broadcasting through YouTube, Ustream, LearnItLive, and with blogs and radio. You can literally run your own TV show, your own news channel, your own broadcasting—right now.

So I encourage you, instead of waiting for the media to discover you, become your own media publishing company. And once you have all of that built out and you're publishing new content on a regular basis, start to contact the media because up until that point you don't have the platform, you're not ready.

I like to say that media doors open best from inside. Be an insider. As a consistent broadcaster of new content, you can, over time, create the power of a personal platform. Others are always seeking a platform upon which to stand with their inspired message.

Develop Camera Presence

Another key point I've observed from working with so many publicists and being a media person myself, is this: Even if your message is incredible and your wisdom is outstanding, if you don't have enough experience being on camera, they're going to pass you by in favor of somebody else who already has a confident bright camera presence.

Invest in yourself. The greatest investment that needs to happen at the very beginning is to invest in your own advancement as someone who can stand in front of the camera and, on all fronts, deliver your message professionally.

Buy a video camera. I strongly recommend that you get yourself a video camera on a tripod and begin standing in front of that lens, looking into it and talking. Talk, talk, talk. Jump into the action and practice expressing your message naturally and powerfully.

Practice. And then input that video content into your computer and just sit there and watch yourself. Don't judge yourself, just observe. You'll naturally begin to see where you can improve your delivery.

Then, get in front of the camera and begin talking again. At first it won't matter what you're saying. What matters is that you learn to feel confident and comfortable as you talk into the camera. Become at home in this new role. Then you can step by step fine-tune what you say.

But notice that you can make a fool of yourself in private, so that by the time you're broadcasting your message you're a bit of a pro, and even enjoy being in the limelight.

Developing personal professional presence is critical because that's your core asset, and it is something you develop by practice and by taking the risk of making a fool of yourself.

If you're an author, get familiar with the role of looking into the camera and talking about your book, about your theme, about your passion. Sometimes work from a script or read from your book until you have repeated the same lines so often that you can say them without the script—and convey sincerity!

Practice does make perfect. Once you've said your basic message enough times, you'll find that you can be in front of the camera and talk spontaneously without a script and express your passion and your information with infinite variations on the basic theme.

This is when you know you're ready for the big time, when you have practiced alone or with friends to where you've fully internalized your basic message, learned it by heart, and can say it in your sleep.

Your Personal Brand

There are many layers to the visual presentation that establishes your brand and personal image. Here's a short list to make sure you pay attention to and harmonize with each layer in order to generate a sense of unified branding:

Create professional photos. You absolutely must have professional photos taken. You'll want at least half a dozen different photos for different media presentations and for variety.

Create high-quality videos. You must have a number of high-quality videos about who you are and what your brand, product, or service are all about.

Use professional design. Professional design can make a tremendous difference. Especially if you're an author of a book, make sure you have a professionally designed book cover.

Become a brand. Make sure that everything you do is true to your voice. Your materials need to be designed as part of one cohesive brand.

Create a website that is easy to use. Design your website around your voice with a good logo and graphics that reflect your brand voice. Make sure that it is clear, clean, and easy to navigate.

Be consistent. Everything that you present to the media and the public must be consistent so that people identify and recognize your one brand, not ten variations on your brand. Make sure that your brand is visually simple, memorable, and crystal clear.

Realistically, you're going to need invest in yourself and hire pro-fessionals to accomplish all of the above, or even a bit more. But without this investment in your brand's future, you can't even begin to play in the media game.

The reality is this—you're already investing loads of your pre-cious time and also your life energy into your product or service. And yes, you will also need to invest a bit of your hard-earned capital. Setting up to broadcast is remarkably inexpensive these days, but still you do need to believe enough in yourself to put some money into your success equation.

Bloom Where You're Planted

Naturally we all want to jump into the big-time regional or national limelight with our media presentations. But the wise move is almost always to test the waters, experiment locally, and build momentum in your community.

If you live in Cincinnati, first attract the news in Cincinnati. Go on Facebook and become friends with the reporters at the local stations in Cincinnati. Watch the newscast often and see what kind of stories they cover. Then send your story ideas into them through an e-mail pitch.

Also, send out press releases to your local stations for events that you have going, and be persistent about it. Call the TV sta-tion and find out the name of the assignment editor.

These beginning efforts to get to know your local news-people don't take very long and don't cost any money or any time. Let's think about this for a moment. When was the last time you received mail at your house addressed to "resident" and opened it? The newsrooms are the same. If you're going to send out a press release to a station's assignment desk, you need to know who's running the assignment desk.

Google+ is an incredible way to connect to reporters. A lot of reporters have a circle live in the newsroom, and they bring people from all over to comment on Google+. Also, LinkedIn is a great way to connect with the reporters and media people in your area.

Continually reach out to expand a general sense of awareness of your topic, awareness of your brand, awareness of yourself, everything about your passion and product. Be your brand, live your message, and broadcast your vision.

Any publicity that you can get has value. You must have content out there if you want to plug in to the media stream of consciousness.

Hold in mind that the media is a hungry beast. They're always avidly looking for high quality content. So if you publish regularly through articles and your blog, and then also send that out in press releases, you will find that if you have something that is needed, interesting, and passionate, you do get picked up.

Go ahead and get your feet wet, especially if you're just starting out. Get yourself ready on all fronts, then send out local press releases; find out who runs your local magazine. Have people write some articles about you to practice your interviewing skills.

Find out who your local news reporters are, and see if you can get on a local show. Perhaps you can piggyback on an event that's related to your topic, or create an event yourself and invite the media to come and attend it.

What you want to do is gain traction, and develop more and more momentum. If you have zero momentum, then you're caught in inertia mode, and you'll have to exert some energy in order to overcome the inertia posture of sitting at home hoping to be discovered.

And remember that there's this powerful machine living inside all the newsrooms in radio, TV, and some magazines—the AP Wire, which stands for the Associated Press. As soon as you present a compelling, burning hot-topic story or are part of one at a local level, you have a high chance of becoming an AP story that can quickly spread like wildfire all over the country, just by pitching locally.

Get Prepared for Television

You might find it very helpful to do the following exercise a number of times, even before you get the camera set up and begin practicing in front of the actual lens. After reading this instruction, go ahead and close your eyes … tune into your breathing and your whole-body presence.

Imagine that you're standing in front of a video camera … stay in touch with your breaths coming and going so that you keep tuned into your emotions and whole-body presence...

Without saying anything out loud, begin talking about your product, your theme, your passion, as if the camera were someone live who is very interested and wants to hear what you have to offer.

Continue talking about anything related to your theme for as long as the flow takes you, and then when you're done, reflect on how it felt to be in front of the camera talking about your passion.

Use the timer on your phone to track how long it takes for you to get the point and deliver a clear, concise, and compelling message.

Pause and Experience

If you want additional support in this area, go to page 203 for instructions to get a FREE BONUS to help your on-camera performance.

"Be yourself;

*everyone else is already **taken**."*

Oscar Wilde

Finding Your
Authentic Voice

Authenticity is an inner quality of your being that emerges naturally and cannot be forced, or faked, or manufactured on demand. Your authenticity is built around your entire life experience, which is then expressed through your life story. First of all, you must of course live your life, learn your lessons, and develop your expertise through direct participation in the field you want to stand out in.

Your Story of Struggle and Triumph

Nothing can take the place of your actual life experience, of your reflection upon that experience, and your integration of that experience into a solid program, product, or service. You gotta walk the walk, and then you can talk the talk.

Many of us have had journeys in our lifetime that have been quite painful. Maybe we're ashamed of our past experience, maybe we were abused in a certain way, but that experience was the key ignition or the spark of our expertise. It was that experience that provoked us to do something about a situation, to fight for a worthy cause, to generate a solution to a problem.

Sharing your story. When you can share the story of your struggle and ultimate conquest in a way that ignites a spark for others, that story becomes not only memorable, it also becomes magnetic. And when it's spoken with a voice of authenticity and not anguish, people can take it in and respond to it and be inspired by it.

This is a big point. If your wound is too fresh, if you've just been through a painful journey and your personal pain is so fresh that you are emotionally changed about it, then all that you will evoke from others will be compassion.

Perhaps an emotion-laden story will cause listeners to say, "Oh boy, I can relate to that. I feel for you." But that's not the response you want. You want them to be inspired by your story and make the connection about how relates to their own life and situation. You want to teach them something about themselves through relating your story, something that they immediately attach to their own life.

In order for you to be a guide or teacher, you need to have transcended your emotional attachment to your story, so that you can say with an authentic voice, "I learned this and I'm going to apply it to you." That's what makes you an expert, that's what makes your story both memorable and inspirational.

The question is, how do you give a taste of who you truly are when talking with someone? It starts with this deep core awareness, you have a powerful story to tell, your are an expert based on your life experience and now you have a clear system that can support others on a journey similar to yours. And this circles back to knowing yourself and being your true self in that moment.

You do have a powerful story to tell, you are an expert in your chosen field. You definitely have a solid program or product or service to offer, to help the person you're talking with. You can relax and be yourself, and always remember that the conversation you're having isn't about you. Your story must be a catalyst that aims attention back to the person you're talking with.

Remember that you want your story to be short, and it ends with a question that returns the focus of attention to your potential client. Return as soon as you can to being a good listener—that's key to being an authentic teacher.

This is where, again, the spiritual journey is most important. You need to be grounded in your core inner vision, and then communicate that deeper sense of who you are, and clarify what space you hold for other people by being a good listener.

Most people will remember you and value you in their lives especially because they get a sense that you really heard them. In this light, often it's not what you say to people, it's what you don't say. It's the space you open up for them to be heard.

Speak The Truth As You See It

Truth. Voicing the truth is always important—that's what makes you authentic, speaking your own unique perception of the reality of a situation. But in this regard, I recommend that you also hold in mind that, as an expert, you are not expected to know the whole truth of life. You are always a work in progress.

Throughout your life you are growing and constantly learning through new personal experience and through regular reflection upon that experience.

Be humble. There must be a certain inner quality of humility involved in presenting yourself as an expert. Your personal presence must express the opposite of an ego trip. People don't expect you to be a know-it-all, all they expect is that you are further along the path, several steps ahead of them, participating in the same world they're in, and therefore able to guide them toward a better life.

Be a teacher. What's important is that at some point you realize that you can stand comfortably in the role of teacher. Ultimately this is an inner decision, a personal act of will, not any kind of external achievement. You simply arrive at the point in your development and awakening where you yourself feel that you are competent and prepared to serve others as a guide.

Be authentic. When you can calmly lead others through territory you know well, you are ready to step into your power, and you can ask for financial remuneration for your services.

A lot of otherwise ready people have difficulty stepping forward and feeling adequate for leadership roles. Some people will never feel comfortable, will never decide that they've arrived, but who's to say when you've arrived? Who's judging, really? And if you require of yourself that you be totally perfect before you claim to be an expert, then you're clearly self-sabotaging your life.

Keep learning. We talked about this earlier, about always being a beginner, and allowing your journey to evolve naturally all your life. It's important to be comfortable with the fact that with each new moment you're learning more about yourself and what you're contributing to the global conversation.

Your expert status in these rapidly changing times is grounded in your ability to participate as an active learner and teacher at the same time in the global conversation.

You don't need to be the loudest voice; you just need to perceive yourself as an active participant. Each day, your understanding of the truth deepens. And when you express your personal view of reality honestly, humbly, and openly, your voice expresses your authentic truth.

Vibrational Credibility

From both the point of view of recent scientific studies and the ancient wisdom of the world meditation tradition, each of us broadcasts our unique vibrational presence out into the world. People we're talking to are naturally impacted by this subtle broadcast of vibrational presence, whether they're consciously aware of it or not.

When you can calmly lead others through a territory you know well, your credibility is accentuated. Your authentic nature creates an unspoken energy field of vibrational credibility. This unspoken wavelength is perceive by others and will determine if an individual will follow your guidance.

Broadcasting negative energy. When you're still caught up in an emotional charge about your life story, you're going to broadcast that vibratory charge outward and hit people with it. They're going to naturally react by pulling back from you, even as you attempt to make them a connection.

Broadcasting an attracting energy. However, when you've taken the time to process and integrate your past traumas and are feeling fairly clear in your own emotions, you'll broadcast an attracting energy. When you come to feel positive and passionate about your story and about the gift you're offering the world, you will broadcast that desired quality of vibrational authenticity.

Radiating likability. When you truly believe in something, when your energy is flowing in alignment with the truth of that belief, you'll find that you radiate an instant likability and also an instant implied credibility.

As an example, I met a potential client the other day and said, "I am a coach for authors in the spirituality marketplace." I said this sincerely and confidently, speaking my personal truth without pushing. I said one sentence to her, and she connected immediately to my vibrational credibility. She knew that I was there for her and could be of help and agreed that same day to work with me as a private coaching client.

Set Limits And Thrive

As I said before, often the less said the better. Radiate your confidence, expertise, and passion. Speak your truth, and people around you will connect immediately. Be sure not to claim to be an expert in too wide a field. Be a specialist in what you are most confident in. Limit yourself to helping people where you feel you will have major success. This is a very wise path to follow.

Present yourself as an expert in one specific area. You'll find that a focused niche, with clarity of purpose and competence, will open all kinds of doorways and bring in more income than claiming to be more than you realistically are.

Always hold in mind that an important aspect of your success as a enlightened expert will be your performance record. Perhaps no one right now is studying the effectiveness of your process or product scientifically, but you can rest assured that, by word of mouth, you will be rated according to the perceived impact of your service.

Especially in the Internet world, word gets out very fast if you're a highly successful guide, and in the opposite direction, word spreads like fire if you're claiming to be able to do something and not delivering the goods. Make sure you deliver what you say and make your voice is a valuable voice of truth.

Voice Activation
Exercise #10

Vibrational Credibility

Take a few minutes and reflect on what we've been talking about. Tune in to your breathing, relax, and look inward and perhaps write down what you feel is your expert arena.

Take an Inventory:

- *What have you learned in your life?*

- *How can you truly help others end their personal suffering? What motivates you to be the recognized expert?*

- *What would be the most incredible experience to happen to your client?*

- *Look honestly at your answers. Are you in alignment? Does your mind, your intentions, and your soul's desire fall into a clear direction of flowing energy, or is there internal conflict?*

Pause and Experience

"When we love, we always strive to

become better than we are.

When we strive to become better than we

*are, **everything around us***

becomes better, too."

Paulo Coelho

Branding Your Expert Voice

Many people develop a great product or service, and then work very hard to push their particular contribution out into the marketplace. But they make the traditional mistake of focusing almost completely on the sales of their books and products, instead of focusing on their own identity and the "sale" of themselves.

You are the Brand

Branding is at the core of your success, but most don't understand how to build their brand correctly. For instance, let's take books—so many people just put their book out front and expect the fact that they're published to somehow work miracles for their career.

Think of any book that's out there with at least a mild amount of success. In today's marketplace, there are 20, 30, 40 or 50 books on the same subject, maybe hundreds or even thousands of books on the same topic. Most of them sell hardly any copies at all, and so the author assumes that he or she is a failure.

Then what do you do? Write another book? Too many people are always starting over because their product didn't sell. They need to realize that products don't sell themselves.

There is almost always a buzz, a story, a personality, a strategic plan that's bringing the product before the public in such a way that a brand becomes born.

Build your brand around your name. However, when you build a brand not around a particular product, but around your own name, you are never starting over. You are always present and ready to advance. You are just adding another product that you are going to try, or another book, or another piece of your content.

You need to think of your book as your calling card. It is a vehicle for boosting your status as an expert in your field.

There is no doubt that a book gives you tremendous leverage

and credibility. It is a fantastic business card, and I personally think every expert can benefit strongly from creating a book. But the fantasy out there in the author world that you can live and survive on just a book—this false hope needs to be put aside.

Rather, your strategy needs to be multi-pronged. People in the expert industry make most of their money through a combination of coaching, live seminars, products, books, and speaker fees. There is a whole industry around this. Successful enlightened celebrities are authors, speakers, and coaches.

If you desire to enter the expert industry, yes, I encourage you to have a book, but don't make your book the cornerstone of your strategy. Instead, make your own presence the cornerstone, the foundation of your growing platform. It's this personality-driven, multi-pronged approach that you need to develop.

Think for a moment about the big celebrities, people like Johnny Depp and Brad Pitt. They play different roles all of the time. But we still know them as who they are, beyond their particular roles.

You need to see yourself in the same light. People respect and know your books and products, but they also are aware of your ongoing presence.

Creating Your Brand Style Guide

One of the most important actions you can take is that of identifying a particular message that you stand for. What is your prime passion that resonates with what people genuinely need in the world? Boil that passion and message down to a one-liner that you feel represents who you really are as an expert.

Next, pick a certain look that represents you. For instance, choose one picture that really exemplifies who you are and use it consistently front and center on all your platforms. This is called your style guide. Use it over and over again with every new representation of yourself. It has the following elements:

Colors: Identify which colors best capture the vibration of the energy or the feeling of your brand.

Name: Decide what name or label you want to be known as or identified with.

Core message: Select your core message which is either your tag line or a full sentence.

Keywords: Determine the five or ten keywords that you want to dominate on the web.

Articles: Make a list of the ten articles you're going to write on your blog.

Videos: Choose the main video that you're going to talk about as best representing who you are.

Consistency: Make sure that everything is congruent and consistent, all the way through your media presence.

If you do this correctly, and present your identity consistently over time, then when anybody encounters anything with your personal brand, even before they see the name, they begin to recognize that it's you.

Focus on Long-Term Payoffs

It takes time to establish a brand; this is your life investment. When you work steadily to achieve consistency, authenticity, and congruency in your personal brand, this raises you above your individual products and books and everything else, and helps you to develop more lasting relationships with both your clients and the media.

The reality here is that you want the media, and the world, to identify you instantly and without hesitation as this authentic true personality that you are, as an ever-emerging authority, and not limited to a particular book that you've written or product you've developed.

A good, honest, revealing book certainly helps. It presents your story and your incredible journey, almost like writing your memoir. Hold in mind, for all expressions of your story, that most people want to create in their minds a strong believable myth about your greatness because then, as they identify with your story and begin using your product or process or whatever in their lives, they feel deeper trust in that process or product.

And in business, what drives long-term success is of course the phenomenon of multiple purchases or involvement in your product/service line. It's so easy to be short-sighted and think of only the immediate payoff. But in reality, your present-moment actions not only sell your services today, they lay the foundation for many more sales to come as you establish lasting brand recognition and value.

What truly captures people's hearts and imaginations is the person behind the story, the real-life character behind the book. It is not the actual words or the book cover or any of the mechanics that present your story; it is your true living essence and unique expression that makes you so magnetic.

From my observation, in the large majority of cases it takes years to develop a brand. There are so many success stories of people, actors as well as everyone else, who struggled for a decade or more before "taking off" in the media and becoming a star. Think of Kevin Costner, a classic example, and Jim Carrey. They were in Hollywood for a long time before becoming famous.

In all honesty, you don't know if it's going to take ten days or a decade to achieve your goal as an expert in your field. Luck is such a strong and uncontrollable variable, not to mention timing and all the unconscious factors that influence your progression to success.

All you know right now is that no matter how long it takes, you want to become known as the go-to person for one particular area of expertise. So pick carefully what that thing or solution is for you. Identify with something you're really passionate about, because later on it can be hard to switch.

Brands Create Media Opportunity

As I mentioned earlier, now more than ever it is important to set the goal of becoming your own media company by using all the traditional and new outlets available to distribute your unique message in a dynamic and magnetic way.

This is equally true for business owners, professionals, entrepreneurs, and others. You must create and broadcast a professional series of content, mainly video and written, that clearly illustrates the way you uniquely solve the problems of your ideal customers.

You can own your niche. The person who steps forward in the media and is perceived as offering a new solution to a source of high-priority pain usually attracts the majority of the marketplace inside that niche.

For example, if you are a divorce lawyer and you create content around peaceful rapid resolution delivered with fairness and equality, you will attract the attention of individuals who want to avoid a painful drawn-out divorce that consumes a large amount of the marital assets.

This formula can be applied to any business or platform. The idea is to create stories and testimonials on video that provide social proof that your business, book, product, or service does live up to the promise it states.

You can become your own media channel. The media is expanding so rapidly and now you have the opportunity to compete directly with mainstream television using Google TV and Amazon TV as platforms. This will revolutionize the integration of online content and traditional television broadcasting.

If as an expert you strategically advance to where you are well-prepared production-wise for this amazing opportunity, you will be able to position yourself as a branded celebrity expert in mainstream television. But you'll need to start right away, if you haven't already, producing your videos in high-definition quality.

Right now only 20% of the content on YouTube is available in HD, even though the technology of the new TVs is becoming increasingly more crisp and higher resolution. Imagine what your low-resolution videos will look like on a 50-inch high-definition television. This isn't the expert image that you want to portray about your message.

Here is a good example of someone who is systematically building a media expert platform. Henry spent the majority of his business life working for other people, but at a certain point he wanted to escape the corporate trap and go forward on his own. With a family to support, he knew he would have to advance step by step toward his dream of independence, so he created a system where he spent just one hour a day working to achieve his goal of being an entrepreneur.

Within a few years, Henry was able to build a six-figure business—in just one hour a day. He then quit his day job entirely, and wrote the best-selling book, *The Hour-A-Day Entrepreneur*, telling his story. He has now built a studio in his home to broadcast a weekly TV show in full HD.

Henry is a perfect example of someone who has learned to successfully broadcast his vision, who has found the money in his message. I'm proud to say that he attended my training program for doing just that, and using some of my video production and marketing strategies, he's created an entrepreneurial business that's already generated over $1 million in revenue.

Fine Points On Developing Story

We've talked about the importance of story throughout this book. This is where so many people flounder rather than succeed as they try to put together a business promoting their expertise. Let's talk one final time about your story.

Pause and ask yourself these key questions about your story:

- *Is it compelling?*

- *Can you say it in two minutes?*

- *Is it genuine?*

- *Does it speak your passion?*

- *Does it express a common problem?*

- *Is your solution believable?*

- *Can people repeat it easily?*

Make it vivid. Have you ever noticed that when you are told a story, the stronger visual images of that story linger within you for quite some time. So be sure that you provide visual elements in your story, images that stick in the mind and help people to re-tell your story so that it spreads virally.

Make it resonate. When people hear a story that resonates with events or situations or problems in their own life, they often remember the story for years. Even though the name of the person and all of the other content around the story is quickly lost, certain images and realizations continue in their minds. Pay attention to what particulars seem to attract your listeners most deeply, and highlight them in your formal story.

Make it transformational. Remember that a good durable story invites others to participate vicariously in the transformation of a person's life experience without their having to go through the pain of transition themselves.

This type of vicarious learning is very important to convey, and you will want to set up your story properly to achieve desired results—telling each stage of your life experience in sequence so that your listeners or viewing audience can put themselves into each part of the progress, from pain to realization and healing.

Make it sell your value. In your story, make sure that the climax highlights your process or service clearly. The point of telling the story is to show how you learned to relieve your own pain. You want to become identified not only as someone who's suffered as others have, but also someone who has achieved relief and realization through your unique but also universal journey.

This primary story of yours will become a most important part of the content that you broadcast to build your platform. Take seriously the challenge of mastering the new media that will enable you to present your story vividly and convincingly.

Let Truth Lead You

However, don't fall into the so-common trap of thinking that you must come up with a super-dramatic, mostly fictitious tale in order to be successful. Most people when listening can tell, at least subconsciously, when you wander away from truth into make-believe in your story, and there is nothing that destroys credibility faster than falsifying your own story.

The temptation is of course very high. You want to be as impressive as possible. But in the long run it's a fatal error to present yourself as something that you are not. Everything rests on credibility, so stay humble, be yourself, and at the same time, give yourself room to be great.

What sells most strongly is passion grounded in truth. You must believe in yourself first, and you can only believe in yourself if you're being authentic.

Voice Activation
Exercise #11

Advancing Your Story

To end this section on becoming master of your own story, I'd like to give you an exercise in advancing your story, an inner process that you can move through regularly in the months to come because your story will indeed evolve as you yourself evolve.

As with the other exercises in this book, set aside some free time, perhaps five to ten minutes. Relax, get comfortable, and quiet your mind momentarily by tuning into your breathing—the air flowing in and out of your nose, the movements in your torso as you inhale and then exhale completely …

Now simply be aware of your body's physical presence here and now, experience your posture, the sensations on your skin— and "be" yourself as a physical presence. This is the body that has moved through all your life drama. This is the living organism that has felt everything that makes up your life story. So always drop back into this physical base of who you are when approaching your story …

Now, focus your attention, while remaining aware of your breathing and physical presence, on your inner passion to be of service in the world. *Feel your physical drive to get out and work with people as you deliver your service or product ...*

As you stay tuned into the actual actions of service that you want to offer your clients, let your mind move back to a starting point in your life story *where something happened that awakened the flow in your life that has led you to where you are right now. See what memory comes to mind as the beginning point of your story ...*

Now see what the next linear step in your story is, into the second part of your story ... *continue moving forward with your story until you've remembered the primary dramatic experiences that have made you who you are, related to the product or service that you're offering the world ...*

And as you finish this inner progression of associations through your story, relax and establish in your mind each of the steps in your story. *See how many steps you have in your story, and then reflect on which three or four steps are most important.*

Which memories carry the most impact on listeners?

Which memories compress your message the most so that it can be told in short form when needed?

Now as you relax with your primary memories vivid and almost alive inside your body, just breathe into the feelings in your body, the actual living story that you contain within you, and that you can tell to others because you can re-experience the story yourself …

Do this process often in the next weeks. This is one of the most powerful ways to become fully genuine and magnetic when you tell your story!

Pause and Experience

"I alone cannot change the world,

but I can cast a stone across the waters

*to **create many ripples**."*

Mother Teresa

VoiceRipples™

September 2012, I found myself standing in the TV news-**room** at the ABC affiliate in St. Louis Missouri. More than 10 years had passed since I last set foot on this property. Previously, every morning my alarm would rattle at 3:30 am. During this time, I was a young mother with a 10-month old daughter. I did not dare to hit the snooze button more than once for fear of permanently falling back to sleep until sunrise. Often, I would only have a few hours of sleep because my daughter kept me awake with her frequent ear infections.

I used to pull myself out of the warm covers and walk down to the bathroom in the guest room so I would not awaken my sleeping husband. Every morning, at 3:30 am I applied a full face of make-up in preparation for my morning media broadcast. Once I was dressed, while it was still dark, I drove 30 minutes towards the station to prepare the morning news for broadcast. This happened every week, Monday through Friday, and once the news was completed in the morning I would jump in the news truck with my videographer named George Wise where we would travel around the metropolitan area of St. Louis in search of a story for the evening news.

My Amazing Opportunity

Now I'm looking at the same newsroom, currently it is empty, a sign of the times. The newsroom shut down ten years ago because of its inability to sustain itself through advertising sales. News is an expensive program for a TV station to run. When a community loses a large number of viewers to online platforms, they also lose revenue from advertisers. Across the country today, many TV stations have already lost their news departments, and all are at risk of shutting down in the near future.

This situation created a unique opening. "Would you like to use the studio?" asked the general manager. "Yes!" I responded immediately.

I took a look around the abandoned studio. It did not look like what one would expect an ABC television studio to appear, evidence that the world of media had permanently shifted. No one had set foot in it for years. Folding tables were haphazardly arranged on the former news set covered with red-and-white-checkered plastic table cloths. In the light grid above my head, many of the lights were rusted and burned out, and a 10-foot long piece a fake, green, holiday garland hung from the metal frame supporting the lights. The carpet was stained, several of the desks were broken, and the rooms were littered with dozens of pieces of dated equipment.

Renovation. In deciding to use the TV studio, I was committing to a series of massive renovations. Several years earlier, I renovated my house, which was built in 1866. It took four years and contractors lived inside my personal space. I knew from experience that renovating the TV station would have surprises and unexpected setbacks. I was right, but I still liked the idea.

Serving emerging experts. Over the past several years I had created hundreds of videos for business owners, authors, speakers, and coaches. Each one had a powerful message that no one was aware of. They were what I called emerging experts. *Emerging experts all have one thing in common: They are lacking an audience.*

I realized that each of these inspiring individuals needed a platform, so I decided to use my expertise in TV news and create a show that would allow me to showcase experts and their content. Many of these people had never been on camera before. They did not understand the importance of creating a precise media-ready message that instantly connects the emotions of the listener. By featuring them as expert guest on the show, they could learn in a real-world environment and teach others at the same time.

My giant leap of faith. This required a giant leap of faith, a 100% committed effort of all of my resources, time, energy, and money. The opportunity presented itself. I knew it would only be offered once. On an intuitive level, I knew the answer was yes, even though I was still unclear about the how.

The following weekend I reached out to my community. This is a clear example of the importance of building an e-mail list of the individuals who support your philosophy. When you have something important to offer, people are already listening to you. I sent an e-mail to my community with the headline, "Looking for 4 Experts to Appear on ABC TV Pilot." This led to a large influx of applicants.

I spent the entire weekend, including the day of my birthday, speaking to people who had been following me for years. By the end of the weekend, I had booked enough guests willing to invest money to go through my media training program to completely renovate the station and start the show as an independent producer and distributor.

Let's start a ripple! The show was called, *Kristen White, Let's Start a Ripple*. Using the media, the Internet, and my voice, I made an impact and started a ripple in less than week.

What could you do in a week with your vision if you dared to take action?

To start, I negotiated with the TV station to air the program in six US TV markets every Sunday to 20 million TV households once the shows were produced, edited, and "in the can." Then I invited my expert guests to meet me in St. Louis for a production day in my newly renovated studio.

Manifest your destiny. It's interesting, let's take a moment and point out a parallel, St. Louis was the gateway for many western settlers during the time of Manifest Destiny in the 1850s. Now, I believe, St. Louis is the launching pad for what I call *Manifest Your Destiny*, in the massive expansion online using a multimedia platform to found an inspirational city.

Like the earlier settlers, I've had a fair share of challenges, times of famine as cash flow crunches, and unexpected obstacles with people who I thought would guide me the right way, but instead created almost fatal delays. However, by divine grace we have made it and continue to grow and expand our territory.

Your Amazing Opportunity

You are personally invited to join me and walk alongside me on my journey. Learn what you need and head off on your own. I'm writing this book to teach you everything I know so that you can succeed and thrive as a messenger. My platform is Transformational TV. I want to help you make your transformation.

When I was in the news, it was polluted with negativity. Everything was about crime and emotional distress. Prior to working in St. Louis, I was a reporter in West Palm Beach, Florida for both the CBS and NBC affiliates. At the time, I had an agent who represented my work. I frequently sent her tapes of my recent stories so she could find me a job in a larger market. Every time I would mail her a tape, she would call me within a few days and request anything other than a story about a homicide. I responded that here in this market, all that I covered was homicide, house fires, and heartbreak. It was a true statement.

A voice for the voiceless. What attracted me to news in the first place was a belief that news is a public service that, at it's highest expression, protects under-represented citizens from exploitation. I used to say that as a reporter I was on a mission to be a voice for the voiceless. Perhaps I was inspired to this calling by the tragedy that struck with my ancestors and claimed my great-grandmother's voice the rest of her life. Whatever the case, I have spent my life in media telling thousands of people's stories with the hope of sharing insight that would protect, educate, and inspire. Let's start a ripple!

Be Your own Media Broadcast Company

Create your media hub. A media hub is the primary location where someone who has been inspired by your content is directed to go so they can instantly take action on what you have offered. This action can be: purchasing your book, scheduling an appointment, or investing in your business on some level.

Your media hub is hosted online, the domain name can use your keywords or the name of your business. For example, if your business is about poison ivy removal, your domain may be www. poisonivyremoval.com or it may be your personal identity.com. In my experience, a media hub based on your name is more effective and enduring.

It's called a media hub because the spokes of all of your multi channel media flow back to one central location. If you are using multi channel media strategies, you will have messages on social, mobile, mainstream media, online radio, and web TV platforms. Each one of these VoiceRipples™ will have a clear emotion-based call to action to go to the URL of your media hub to learn more, access free content, and to take action.

Your Media Hub Components

Videos. All the videos of your previous media appearances should be on your hub. If you have not been on TV before, you need to get at least one media appearance under your belt to generate more media invitations.

Studies have shown that media experts convert ten times more customers than individuals who have not appeared in the media. Take your current sales and multiply that by ten. Now look at the number and imagine what media can do rapidly to the the bottom line of your business.

Media doors open easily from the inside. This is important, because media appearances give you authority and leverage and increase your sales by a multiplier.

Button/link. Next, immediately underneath your media appearance video, there should be a button or a link to a web form for the visitor to take action and engage in your product or service. For example, this can be an opportunity to buy the book, to enter their e-mail in exchange for free content, or to enter their e-mail and phone number to schedule a brief discovery session to learn about your consulting services.

It is also possible to add the technology of a live chat to your website to capture and engage people at the peak point of interest.

Please note, offering a subscription to a newsletter is not engaging. This offer translates into more e-mail to manage and rarely is it acted upon. If this is your only offer, you absolutely must come up with a new one.

Media kit. In addition to the video of your media appearance and your "buy now" button, you must also create a media kit. In this kit you will host the following content: a professional photo, media-ready questions, a professional bio, a description of your product, business, or services, a high quality photo of your book cover, your book description, and the name and phone number of a person acting as your publicist. The media kit will also include press releases, previous appearances, and any other media coverage.

By hosting this on your website, it makes you media ready at every moment. Media coverage happens when opportunity meets preparation. A reporter will be looking for an expert, visit your website, review your media kit, and then call you for a comment. If any part of this is missing, they will move on quickly to someone else. The world of news moves at lightning speed.

Why do You Need a Media Hub?

The answer is straightforward. You are leaving money and opportunity on the table without one. I have encountered hundreds of people who have been featured in the news media and talk shows dozens of times. Unfortunately, these people have not been able to monetize their media experiences. Why? Because when you are featured in the media, the people watching will only remember you for a moment.

To attract and engage. If you have a clear call to action, and if your message connects to them emotionally, then they will immediately look you up online, visit your media hub, recognize your video, see the call to action, and press the "buy now" button underneath your media credentials. In this set of circumstances, and only under these circumstances, will you be able to generate income from your singular media appearances. Otherwise, you will be featured in the media today and forgotten by tomorrow, because that is the nature of the beast.

With each media appearance, it is important to recognize these key steps:

- *Attract*

- *Engage*

- *Connect*

- *Convert*

Your message should be precise and designed with language that attracts your ideal customer. In the media, you have 15 seconds to attract the attention of the viewer before you have lost their attention completely. Media messages that are designed to attract everyone generally attract no one. Likewise, media messages that speak directly to the pain of a unique niche and provide a solution to this pain, instantly create an emotional bond that leads to the next step—engagement.

Once you have engaged your audience, you have their attention. They are fully present listening to you as the expert outlining specific solutions to their painful and complicated life situations. If you are able to use your voice, to create language that mimics the ongoing dialogue within the mind of your ideal customer, then they will connect to you instantly and form an emotional bond.

To connect and convert. In the next step, connection, the listener will take an action if directed. This action may be to visit your website. Studies have shown that 90% of people today watch TV while using a laptop, smart tablet, or phone. This means online access is instantaneous with the broadcast of your message if an invitation to visit your website/media hub is extended.

Once they connect to your website, you have an opportunity based on your offer to convert. Convert means an interested individual hears your message, visits your website, and enters their name and e-mail address. This sequence is necessary to create a successful relationship with customers using immediate expert platform.

Once you have a name and an e-mail, you can invite individuals to attend webinars, to engage with you via social media, to listen to your other broadcast, to schedule a discovery session, and to take action on a variety of offers that you will extend to them for the duration of your relationship. Many people will stay on your list for years and when your offer is right they will turn into clients.

Use Your Voice to Start a Ripple

The strategy is to send out a powerful, sustainable, consistent, transformational message. I call these messages *VoiceRip-*

ples™. They are powerful ideas that once shared with the public make an impact and spread your philosophy around the globe to everyone who is receptive to receiving your inspiration.

These VoiceRipples™ are best shared in the media because this allows them to spread globally. The media landscape is quite different today. It is constantly changing and there are many channels for you to choose to *VoiceCast*™ (my new term for the evolution of broadcasting your voice message).

As an expert, you need to leverage as many channels as possible concurrently and consistently to create the energy needed for visibility. I call this creating a Multi-Channel Media platform.

Multi-channel media platforms. What are the categories in a multi-channel media platform? I hesitate to share specific channels because even as I write this manuscript, the list of outlets has multiplied. For the purpose of this book, I will divide them into several categories. You can visit our website (listed at the back of the book) for the most up-to-date list.

A multichannel media platform combines the use of these main VoiceCast™ outlets:

- *Mobile*

- *Social*

- *Television*

- **Radio**

- **Publishing**

For example, as a media expert developing an expert platform, you may host a weekly radio show, host a TV show with several episodes, and publish a quarterly magazine. You become you own media company. In addition, you have an outflowing media strategy, where you are a guest on other people's shows, a speaker on their stages, and a columnist in other magazines. Your VoiceRipples™ are sent through a variety of these channels.

The Multi-Media Evolution

Mobile devices. Mobile devices and smart tablets are the fastest growing communication mode on the planet. In a very short time, almost everyone across the globe, even in areas of desperate poverty, will be able to connect and send VoiceRipples™ using their phones. Now, on your phone it is possible create HD video, send this video directly to You Tube, publish content, utiltize innovative technology in apps, and make purchases. This technology will expand and refine rapidly in the near future.

Social media. The megaphone for your voice is social media. With more than 1 billion users, social media is the main tool of recruiting people to follow your movement. A perfect example is a man by the name of Drew Canole. Drew is an expert on juicing vegetables. Now a lot of people out there believe in juicing, but

Drew has created a multi-channel media expert platform.

He has 500,000 followers on one Facebook page and 750,000 followers in another Facebook group, which are all extremely active. He publishes content to his social media platforms every day. Likewise, he posts a video several times a week via his TV channel FitLife TV. And, he is the author of two best-selling books on Amazon. Drew embodies the success that is possible by sustainable content that is inspiring and delivered via the voice of one individual. His is creating daily VoiceRipples™ that are consistently making the waves necessary to build out an inspirational city.

Television. Television has experienced the most revolutionary changes in the past decade. Newscasts are losing viewers. You Tube has earned the designation of the number one search engine. WebTV shows are popping up every day. Google TV and Amazon TV have launched a subscription-based service direct through in-home TV portals. Services like Hulu and on-demand cable allow people to watch movies and series on all of their devices anytime they want.

Radio. Radio is still an amazing media. A perfect example is a radio platform called VividLife. Four years ago, I was one of the first hosts on this new online radio station which featured content on wellness, spirituality, and personal development.

Every week, on my show, I would interview a best-selling author.

As part of the interview protocol, the guest promoted the show, I promoted the show, and VividLife promoted the show.

We all sent out VoiceRipples™ to our communities through a one-hour interview of transformational expert content available for anyone to listen. This created energy and new people to the shared platform. Very quickly, VividLife became viral and grew to become the Best of Blog Talk Radio and is now considered the go-to resource for this genre. Their syndicated blog is now read by millions of subscribers.

Self-Publishing. Books are the new business cards. Any expert, who has a message to share can amplify their credibility and leverage their influence by creating a book about their unique insights and strategies. Imagine going to a meeting and handing out your book in place of a business card. *How would this impact your ability to attract people who resonate with your cause?*

According to Publisher's Weekly, eight out of ten people believe they have a book inside of them. You may be one of these people. The question is, do you have the courage to write it? Your voice chronicled inside a book is a powerful and lasting legacy.

Share Your Message with the World

You can send your VoiceRipples™ across the globe using all of these multimedia channels and start a movement now. It is up to you as the messenger to decide who you are speaking to and when you want to start sharing your voice. There is a journey that unfolds for every thought leader. It takes time, energy, and focus to keep moving forward, but if you have a volition to be heard, I know you will prevail.

The world is waiting to hear what you have to say. Now is your moment to claim your expertise, unlock your voice, and step into the spotlight as a beacon of transformational energy for the planet. ***Let's Start a Ripple!***

Voice Activation
Permission Granted

VoiceRipples™

"What do you have to say?

Who do you want to be?

What are you destined to do?

Life is about the impact you make

with the message you share.

Find your voice...

Let's start a ripple!"

About the Author

Kristen White brings more than a decade of experience as an award-winning television journalist and anchor to her new show, The Ripple Effect, airing in six U.S. television markets on ABC, FOX, CW, MyTV, and on Charter Cable. She is a 2012 recipient of two gold Stevie Awards for Women in Business for a Video of the Year, Mentor/Coach of the Year, and a bronze award for Female Entrepreneur of the Year. Kristen has also been awarded the 2012 Platinum Aurora Awards for Video and a 2012 Gold Aurora award for Video.

Kristen is an award-winning media coach who helps entrepreneurs and businesses become more powerful messengers with content creation and on-camera coaching. She works with clients to create multi-channel media marketing platforms including television, radio, iPad Magazine, and multimedia books. Her company, Creative Catapult Video, is the premier media production company for the wellness and personal development marketplace.

Bonuses

Thank you for reading *Voice*.
As a special gift, I'd like to give
you two free bonus items.
They are available at:
www.expertvoicebook.com/bonus

Your bonuses include:

1. Polish Your On-Camera Image with 22 Tips
   from the Media Pros ($97 value)

2. Learn How to Build a Powerful Business Around Your Voice:
   Find the Money in Your Message ($197 value)

Claim your bonuses at:
www.expertvoicebook.com/bonus

Your bonuses give you the following benefits:

- Turn interested seekers into high-paying raving fans
- Instantly make a lasting connection with your target audience by "speaking their language"
- Discover your passion all over again and enjoy more profits
- Make a major positive difference in the lives of others